£9.95

BASIC ORGANIC CHEMISTRY
SECOND EDITION

BASIC ORGANIC CHEMISTRY
A MECHANISTIC APPROACH

Second Edition

J. M. Tedder, Department of Chemistry, University of St Andrews, Scotland
A. Nechvatal, Department of Chemistry, University of Dundee, Scotland

JOHN WILEY & SONS
Chichester · New York · Brisbane · Toronto · Singapore

Library of Congress Cataloging-in-Publication Data:

Tedder, John M. (John Michael)
  Basic organic chemistry.
  Includes index.
  1. Chemistry, Organic.   I. Nechvatal, Antony.
II. Title.
QD253.T37   1987   547   86–18911

ISBN 0 471 90976 9
ISBN 0 471 90977 7 (pbk.)

British Library Cataloguing in Publication Data:

Tedder, John M.
  Basic organic chemistry : a mechanistic
  approach.—2nd ed.
  1. Chemistry, Organic
  I. Title   II. Nechvatal, A.
  547   QD251.2

ISBN 0 471 90976 9
ISBN 0 471 90977 7 Pbk

Printed and bound in Great Britain

PREFACE TO SECOND EDITION ———

When this book was first introduced almost exactly twenty years ago, organic chemistry was traditionally taught as a list of preparations and properties of different classes of compounds. As a result the student was presented with a huge collection of facts to which a little theory was sometimes added as a piquant sauce to make the indigestible mass a little more palatable. Not surprisingly the teaching of organic chemistry has changed dramatically during the twenty years since this book first appeared. Almost all textbooks of organic chemistry attempt to develop the subject as a logical procession of concepts. In the original volume, arguments developed in one chapter depended on those developed in previous chapters and the book was almost meaningless unless read consecutively.

In addition to the changes that have taken place in the teaching of organic chemistry, the introduction of new material coming under the general title of 'Spectroscopy' has completely revolutionized both the teaching and the practice of organic chemistry. Until the revolution in teaching, much of organic chemistry was presented as a series of facts and these began with descriptions of the physical characteristics (e.g. boiling point, melting point, refractive index, etc.) Although these physical characteristics are important in identifying known compounds, they are almost irrelevant to the understanding of organic chemistry.

In the first edition we deliberately excluded the then new spectroscopic techniques. The use of infrared spectroscopy and, more important, nuclear magnetic resonance spectroscopy have developed so dramatically in the last fifteen years that no student of organic chemistry should be without some ability to use and interpret n.m.r. and i.r. spectra. A new chapter (2) has been introduced which outlines the spectroscopic techniques available to the organic chemist.

We have retained the original descriptions of the types of reaction characteristic of particular molecular structures, but we have completely revised our original presentation. In this new edition the majority of chapters begin with a brief account of the spectroscopic characteristics of the class of molecule we are going to discuss. We believe spectroscopy is now so important that all students starting to learn about the chemistry of carbon compounds must, of necessity, also develop an understanding of the spectroscopic methods of structural determination. The book has been written so that it would be possible to read it omitting the spectroscopy in the first reading, and some teachers may favour this. For our part we believe that spectroscopy and chemical reactivity should be developed simultaneously.

The other less important change we have made is to include an additional chapter in which pictorial orbital theory is described. We believe that students will first master the concept of electron pairs developed out of the Lewis theory of valency. However, students will be meeting so-called 'molecular orbitals' at a very early stage and we have therefore felt it desirable to have a brief outline of pictorial orbital theory so that students continuing to study organic chemistry can fit the groundwork, using Lewis theory, with the more advanced ideas in which pictorial orbital theory is essential.

ACKNOWLEDGEMENT ──────

The infrared spectra are taken from the Aldrich Library of Infrared Spectra, 2nd edition, C. J. Pouchert. The proton magnetic resonance spectra are taken from the Aldrich Library of NMR Spectra, C. J. Pouchert and J. R. Campbell. The ^{13}C magnetic resonance spectra are taken from the American Petroleum Institute Research Project 44, and from *Carbon-13 NMR Spectra*, L. F. Johnson and W. C. Jankowski, Wiley–Interscience, 1972.

CONTENTS

CHAPTER 1

Introduction

Organic chemistry is the study of carbon compounds, and the question immediately arises as to why we separate the chemistry of one element from that of all the other 91 naturally occurring elements. The answer is twofold. Firstly, for every chemical compound known containing no carbon, there are many dozens containing carbon. The second reason is related to the first; what we call life is a highly complex series of chemical processes and these invariably involve compounds of carbon. Further than this, any form of life, wherever it may exist, must involve the chemistry of carbon. Only by utilizing carbon compounds is it possible to build up systems complex enough to support life. It does not matter whether visitors from outer space are like Stephen Spielberger's *E.T.* or the *Black Cloud* of Fred Hoyle, the crucial point is that only by using carbon compounds is it possible to create an organism complex enough to represent what we mean by life. Having established that carbon is unique among the elements the next question is, why is it unique?

For the present discussion we shall make use of Lewis's theory of valency, according to which chemical bonds are formed by atoms sharing electrons to form electron pairs, each atom concerned tending to produce the shell corresponding to the inert gas nearest to it in the periodic table.

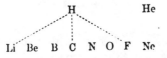

Thus, in the diagram above, which represents the first row of the periodic table, lithium tends to lose an electron to form Li⁺, a

1

lithium cation, in which the lithium nucleus is surrounded by two electrons, as in helium. At the other end of the first row of the elements is fluorine, which tends to gain an electron so that it has eight electrons in its outer shell like the inert gas neon. Let us now compare the compounds formed by hydrogen with lithium, carbon and fluorine. Lithium hydride is an ionic solid best represented as Li^+H^-. It is extremely reactive, almost explosively so, with water:

$$LiH + H_2O \longrightarrow Li^+ + H_2(g) + OH^-$$

Hydrogen fluoride is a low-boiling liquid which also reacts violently with water:

$$HF + H_2O \longrightarrow H_3O^+ + F^-$$

Carbon tetrahydride, CH_4, normally called *methane*, is a gas which is completely immiscible and inert with water. Methane in many ways resembles molecular hydrogen. A molecule of hydrogen does not ionize readily:

$$H:H \xrightarrow{\quad\quad} H^+ + H:^-$$

and it does not react with water. The H—H bond in hydrogen is a typical non-polar covalent bond—the two electrons forming it are shared equally between the two hydrogen atoms—whereas in lithium hydride the hydrogen has a bigger share of the electrons and in hydrogen fluoride the fluorine atom has the bigger share; this can be represented by the simplified picture shown below:

$$Li :H \qquad H:H \qquad F: H$$

Lithium hydride and hydrogen fluoride are polar molecules whereas molecular hydrogen is not. Methane is very similar to molecular hydrogen in that the electrons forming the bonds between the carbon atom and the hydrogen atom are equally shared between the carbon atom and the four hydrogen atoms:

$$\overset{\displaystyle H}{\underset{\displaystyle \ddot{H}}{H:\ddot{C}:H}}$$

Metallic lithium reacts violently with hydrogen fluoride to form hydrogen and the salt lithium fluoride, Li^+F^-. It also reacts violently with water to form hydrogen and lithium hydroxide, Li^+OH^-, which is also a polar salt. Lithium dissolves in liquid ammonia, NH_3, reacting gently to give a salt, lithium amide,

Li$^+$NH$_2^-$, and hydrogen. Lithium does not react with methane at normal temperatures. Notice the decreasing reactivity in this series:

$$Li + HF \longrightarrow Li^+F^- + \tfrac{1}{2}H_2 \quad \text{Very violent}$$
$$Li + H_2O \longrightarrow Li^+OH^- + \tfrac{1}{2}H_2 \quad \text{Rapid}$$
$$Li + NH_3 \longrightarrow Li^+NH_2^- + \tfrac{1}{2}H_2 \quad \text{Slow}$$
$$Li + CH_4 \longrightarrow \text{no reaction at normal temperatures}$$

We have likened the C—H bond in methane to the H—H bond in molecular hydrogen and this suggests that hydrogen could be replaced in a molecule H—X by carbon or, specifically, by carbon surrounded by three hydrogen atoms, CH$_3$ (methyl group), to produce a further molecule CH$_3$—X. Table 1.1 shows such a parallelism between some inorganic compounds H—X and their methyl analogues (CH$_3$—X.

Table 1.1

	H—X		CH$_3$—X	
Hydrogen chloride (fluoride, bromide, iodide)	H—Cl (F,Br,I)		CH$_3$—Cl (F,Br,I)	Methyl chloride (fluoride, bromide, iodide)
Water	H—OH		CH$_3$—OH	Methanol (methyl alcohol)
Ammonia	H—NH$_2$		CH$_3$—NH$_2$	Methylamine
Hydrogen sulphide	H—SH		CH$_3$—SH	Methyl mercaptan
Methanal (formaldehyde)	H—CHO		CH$_3$—CHO	Ethanal (acetaldehyde)
Methanoic acid (formic acid)	H—COOH		CH$_3$—COOH	Ethanoic acid (acetic acid)
Hydrogen cyanide (Prussic acid)	H—CN		CH$_3$—CN	Methyl cyanide (acetonitrile)
Sulphuric acid	H—OSO$_3$H		CH$_3$—OSO$_3$H	Methyl hydrogen sulphate

This process can be extended further in water and ammonia, since there is more than one hydrogen atom which can be replaced by CH$_3$:

HO—H Water	NH$_3$ Ammonia
|	|
CH$_3$O—H Methanol	CH$_3$NH$_2$ Methylamine
|	|
CH$_3$OCH$_3$ Dimethyl ether	(CH$_3$)$_2$NH Dimethylamine
	|
	(CH$_3$)$_3$N Trimethylamine

Hydrogen attached to carbon can also be replaced by methyl, giving series such as that shown below:

HCHO
Methanal
(formaldehyde)

|

CH_3CHO
Ethanal
(acetaldehyde)

|

CH_3COCH_3
Propanone
(acetone)

HCOOH
Methanoic acid
(formic acid)

|

CH_3COOH
Ethanoic acid
(acetic acid)

|

CH_3COOCH_3
Methyl ethanoate
(methyl acetate)

What about the hydrogen atoms in methane? Clearly we can do just the same as illustrated in the scheme below:

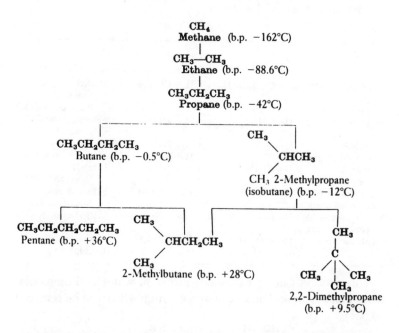

Two molecules with the same molecular formula but different structures are called *isomers*. How many isomers of hexane (C_6H_{14}) are there (cf. Problem 1 and its answer)?

This process can be continued *ad nauseam*, e.g. the common plastic polythene (sometimes called 'Alkathene') consists of very long chains of carbon atoms surrounded by hydrogens, i.e.

$H(CH_2)_nH$, where n is a very large number. The first four compounds in the above scheme are gases; the pentanes are low-boiling colourless liquids and the longer the carbon chain the higher the boiling point. When the chain contains sixteen carbon atoms in a row the boiling point has reached 280° C and the hydrocarbon is no longer a liquid but a low-melting solid (or wax). Similarly, hydrogen chloride can yield not only methyl chloride but a whole series of chlorides with carbon chains attached to the chlorine atom; similarly from water a series of alcohols and ethers can be built up by successive replacement of hydrogen atoms by methyl groups, as shown in the next scheme:

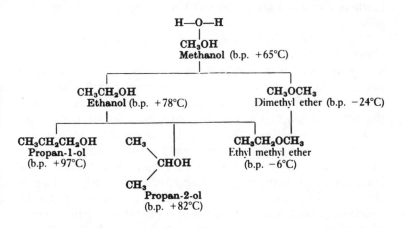

Such series are called *homologous series* (C_nH_{2n+2}, the alkanes; $C_nH_{2n+1}OH$, the alcohols; $C_nH_{2n+1}CO_2H$, the carboxylic acids, are all general members of homologous series where n can be any integer). In all homologous series there is a gradation of physical properties, and in general the boiling points and melting points increase with increasing chain length. Usually a compound with a branched chain has a lower boiling point than its straight-chain isomer.

Notice the duplication of names for some of the simple members of the homologous series. Many of these compounds have been known from the beginning of organic chemistry as products of living organisms—acetic acid, the main constituent of vinegar (*acetum*, Latin for vinegar); formic acid, a constituent of the defence spray produced by some ants (*formica*, Latin for an ant). Systematic names (see the nomenclature section at the end of each chapter) are designed so that the same prefix is used for the same group throughout the variation of structure. All one-carbon derivatives use 'meth':

CH$_4$	Methane
CH$_3$OH	Methanol in place of methyl alcohol
HCO$_2$H	Methanoic acid rather than formic acid

The disadvantage of systematic names is that the unfamiliar systematic name disguises the familiar chemical; ethyl ethanoate, for example, although it tells you precisely the structure of the ester, is a bit of a mouthful and will never replace ethyl acetate, familiar not only in the organic chemistry laboratory as a pleasant smelling liquid but known by this name in industry as a valuable solvent.

Thus we can build up long straight chains of carbon atoms or, alternatively, branched chains with as many branches as we like. The next question is, is it possible to make a chain bite its own tail, i.e. can there be cyclic chains of carbon atoms? The answer is 'yes', but before considering this we must go back and consider the actual shape of the methane molecule. So far it has been drawn as though it were flat, with the carbon atom in the centre linked to four hydrogen atoms, making the whole molecule a square. This has only been done for convenience in drawing. Since the bonds between carbon and hydrogen consist of pairs of electrons, and since electrons repel each other, each of the bonds would be expected to be as far away from the other three bonds as possible in methane. This will occur when the bonds from carbon to hydrogen subtend an angle of 109° 28' to each other. This represents the tetrahedral distribution, i.e. each hydrogen atom can be regarded as being at the apex of a tetrahedron in which the carbon atom is at the centre, as in **1**. Structure **2** represents methane, showing the bonds between the atoms. The tetrahedron has been turned in such a way that the carbon atom and the atoms H$_{(a)}$ and H$_{(b)}$ lie in the plane of the paper. H$_{(c)}$ thus projects out of the plane of the paper and H$_{(d)}$ lies behind the plane of the paper (shown with a dashed-line bond). In subsequent chapters **2** will be represented in the simplified form **3**. It is important, however, that the student should try throughout to visualize the three-dimensional nature of chemistry.

This tetrahedral arrangement of the bonds around the carbon atom means that molecules which we have described above as

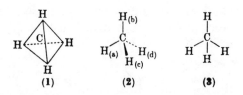

(1)　　　(2)　　　(3)

linear, e.g. the linear hydrocarbons, are, strictly speaking, not linear but zigzag. The direction of the bonds is fixed but free rotation about a bond joining two atoms is possible. Thus, there is only one 1,2-dichloroethane, positional isomer **5**; this can be converted into **4** simply by rotation of the C—C bond. When we now try to make the cyclic hydrocarbons but keep the bonds between the carbon atoms at exactly 109° 28′, we find that only rings of six carbon atoms or larger can be made. Five-, four- and three-membered rings can be made only by sacrificing the tetrahedral angle, the extent of the sacrifice becoming, of course, greater as the ring becomes smaller.

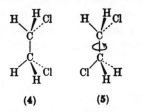

(4)　　　(5)

5 can be converted into 4 by rotation about the carbon–carbon bond
Bonds in the plane of the paper C–H
Bonds behind the plane of the paper
C----H
Bonds in front of the plane of the paper C◄ H

Cyclohexane, C_6H_{12}, has a puckered ring in which the tetrahedral angle is perfectly preserved, and the same is true for any larger ring system. Cyclopentane is almost planar and the bonds are only slightly distorted from the tetrahedral angle. In cyclobutane and cyclopropane, however, distortion is considerable. This introduces what is called *strain*. For the moment we can best interpret this in terms of energy and regard strain as the energy required to move the bond from its tetrahedral angle. This strain energy will have to be taken away from

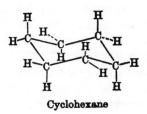

Cyclohexane

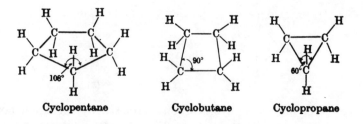

Cyclopentane   Cyclobutane   Cyclopropane

the energy of the bond and therefore a strained bond will be weaker than a non-strained one. Further discussion of cyclic compounds will be postponed until later.

There is one further consequence of the tetrahedral arrangement of the bonds around the carbon atom which must be mentioned here briefly. If we consider a carbon atom attached to four different atoms or groups we see that two arrangements are possible, one of which is a mirror image of the other. These two mirror images, **6** and **7**, are like a left-handed glove and a right-handed glove and are not superimposable on one another. From a chemical point of view, these two compounds will have identical chemical reactions and it suffices here to say that these two isomers, called *stereoisomers*, are distinct species and can be separated by physical methods, though this matter will be discussed in Chapter 16.

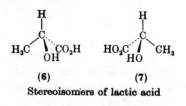

(6)   (7)

Stereoisomers of lactic acid

Nomenclature

We have laid stress on the great number and variety of compounds built up of carbon atoms. It is extremely important, therefore, that we have some systematic method of nomenclature. Such a system has been worked out by a commission appointed by the International Union of Pure and Applied Chemistry. The following are some of the most important rules for the naming of acyclic open-chain hydrocarbons with the general formula C_nH_{2n+2}. The first four such unbranched hydrocarbons are called methane, ethane, propane and butane. Names of the higher members of this series consist of a numerical prefix and a termination '-ane'. Examples of the numerical

Table 1.2 Nomenclature for straight-chain hydrocarbons C_nH_{2n+2}.

$n =$				
1 Methane	11 Undecane	32 Dotriacontane
2 Ethane	12 Dodecane	33 Tritriacontane
3 Propane	13 Tridecane	40 Tetracontane
4 Butane	14 Tetradecane	50 Pentacontane
5 Pentane	20 Eicosane	60 Hexacontane
6 Hexane	21 Heneicosane	70 Heptacontane
7 Heptane	22 Docosane	80 Octacontane
8 Octane	23 Tricosane	90 Nonacontane
9 Nonane	30 Triacontane	100 Hectane
10 Decane	31 Hentriacontane	132 Dotriacontahectan

prefixes are shown in Table 1.2. The generic name for all saturated acyclic hydrocarbons, branched or unbranched, is alkane.

The fundamental concept in the naming of organic compounds is one of substitution. Thus we consider the branched hydrocarbon as being derived from the longest chain in it, substituted by shorter chains. The shorter substituent chains are named as radicals. The radicals are derived from the saturated unbranched hydrocarbon by the removal of a hydrogen atom from the terminal carbon and are named by replacing the ending '-ane' of the name of the hydrocarbon by '-yl', e.g.

CH_3- Methyl from CH_4 methane
$CH_3CH_2CH_2CH_2-$ Butyl from $CH_3CH_2CH_2CH_3$ butane
$CH_3(CH_2)_9CH_2-$ Undecyl from $CH_3(CH_2)_9CH_3$ undecane

We are now in a position to name a branched-chain hydrocarbon. The name of the longest chain present in the molecule is prefixed by the names of the side chains, e.g.

$$CH_3CH_2CHCH_2CH_3$$
$$|$$
$$CH_3$$
Methylpentane

Clearly this name is ambiguous for the following molecule is also methylpentane:

$$CH_3CHCH_2CH_2CH_3$$
$$|$$
$$CH_3$$
Methylpentane

To distinguish between these two molecules, the carbon atoms of the longest chain, that is the five carbon atoms of the pentane molecule, are numbered from one end to the other with arabic numerals, the direction of numbering being chosen so as to give the lowest possible number to the side chain. Thus our two examples become 3-methylpentane and 2-methylpentane:

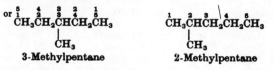

Many small common molecules have a common 'semi-systematic' name as well as the systematic name; 2-methylpentane is also known as isohexane, 'iso' denoting the presence of a $(CH_3)_2C$ group in a hydrocarbon. If two or more side chains of a different nature occur, they are cited in alphabetical order, multiplying prefixes 'di-', 'tri-', etc., being ignored.

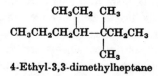

The cyclic hydrocarbons of the general formula C_nH_{2n} are named by prefixing 'cyclo' to the name of the straight-chain compound with the same number of atoms. Some examples are given below:

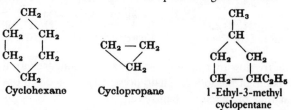

There are additional rules to deal with more complicated molecules but the above are sufficient to deal with any hydrocarbons which we are likely to meet here. We would advise the student to have a little practice at naming some hydrocarbons (cf. Problem 1).

Problems

1. Draw out all the possible isomers of molecular formula C_7H_{16}. Name them.

2. In the text a particular shape (*conformation*) of cyclohexane is shown. Draw the possible conformations of methylcyclohexane, $CH_3C_6H_{11}$, in which the tetrahedral arrangement of carbon atoms is maintained.

CHAPTER 2 ————————————

Structure Determination

The majority of organic molecules are either white crystalline solids or colourless liquids. In the past organic compounds were characterized by their crystalline form and melting point if they were solids and by their boiling point and refractive index if they were liquids. The structures of organic molecules were founded on accurate chemical analyses, followed by a study of the chemical properties of the unknown substance including, if necessary, chemical reactions which degraded the unknown molecule into simpler fragments which could be identified. Chemical combustion analysis, in which the organic molecule is burnt in an atmosphere of oxygen and the resultant water and carbon dioxide are very accurately estimated, is still a very important technique, but it is used to confirm structural assignments and above all to confirm purity of products whose structure has been determined by spectroscopic methods.

The first step in any study of a new organic compound must be its purification; therefore recrystallization and distillation are the first techniques a student meets in the laboratory. As the laboratory work progresses various forms of chromatography will be encountered.

> Column chromatography
> Paper chromatography
> Thin layer chromatograpy and
> Gas chromatography

All these techniques are used to purify organic compounds. The next step is the use of the mass spectrometer to determine the molecular weight of the compound, followed by quantitative estimation of the elements present. Once the molecular formula has been established the nature of the functional groups present in the molecule can be determined by physical observations, especially spectroscopic techniques.

Virtually all the regions of the electromagnetic spectrum, from X-rays to radio waves, are used in the study of organic molecules, but three regions are particularly important. These are the near-ultraviolet (200–400 nm), the infrared (4000–600 cm^{-1}) and then, in conjunction with an applied magnetic field, radio waves (60–400 MHz).

Light Absorption and Energy Changes in Molecules

The total energy of the molecule is made up of binding energy (electronic energy) and kinetic energy (vibrational and rotational energy):

$$E_{total} = E_{electronic} + \underbrace{E_{vibration} + E_{rotation}}_{} \qquad (1)$$

$\underset{\text{energy}}{\text{Binding}} \qquad \text{Kinetic energy}$

All molecular energy is quantized, i.e. a molecule can only take up specific values. The electronic energy levels are relatively far apart; vibrational and rotational levels are successively closer together in energy. When a molecule absorbs a quantum of light the frequency (v) of the light is related to the energy E by the expression:

$$E = hv \qquad (h = \text{Planck's constant})$$

The absorption of ultraviolet radiation results in the promotion of an electron from one electronic level to a higher one. Transitions in molecular vibration and rotation are associated with infrared light. Infrared radiation has far less energy than ultraviolet radiation. For example, a quantum of ultraviolet radiation of 2000 Å or 200 nm* (5×10^4 cm^{-1}) has an energy equivalent to 142 kcal mol^{-1}, while a quantum of infrared radiation of 10 μm (1000 cm^{-1}) is equivalent to only 2.9 kcal mol^{-1}.

---

* 1 nm (nanometre) = 10^9 metre
1 μm (micrometre) = 10^{-6} metre

Ultraviolet Spectroscopy

As the name implies, the ultraviolet region of the spectrum extends from the short-wave limit of the visible spectrum (i.e. the violet) at about 400 nm to the region of soft X-rays at about 100 nm. Not all the parts of this extensive region of the spectrum are equally accessible experimentally. Below 200 nm, air (especially oxygen) and silica are no longer transparent. This region of the spectrum is often called the 'vacuum ultraviolet'. It is, however, the longer-wavelength region (200–400 nm), the near-ultraviolet, which is of greatest interest in organic chemistry. Single bonds, i.e. those bonds which involve the sharing of two electrons by two atoms, only absorb light in the inaccessible vacuum ultraviolet region. When two atoms share four electrons, i.e. are connected by a double bond, light in the near-ultraviolet region is absorbed and this is the region of the ultraviolet spectrum which is of greatest interest in organic chemistry. As well as carbon–carbon double bonds, carbon–oxygen and carbon–nitrogen double bonds also absorb light in the ultraviolet region. The absorption bands of the carbon–oxygen double bond and the carbon–nitrogen double bond are relatively weak but nonetheless can be important in structural determination. When double bonds are joined together (see the discussion on conjugated double bonds in chapter 14) the intensity of the absorption increases and the absorption band is shifted to longer wavelengths (i.e. towards the red), see Figure 2.1.

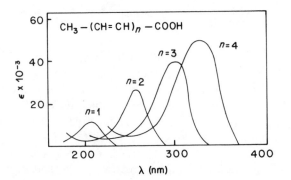

Figure 2.1 Variation in wavelength and intensity of the first absorption band of a polyene acid with chain length. (*Reproduced by permission of Methuen & Co Ltd from* Theory of Electronic Spectra of Organic Molecules, *p. 69.*)

Infrared Spectroscopy

The infrared region extends from the visible region ($\lambda \approx 7 \times 10^{-5}$ cm) until it merges with microwaves ($\lambda = 0.1$ cm). The region of the infrared spectrum most used in organic chemistry ranges from about 2.5×10^{-4} to 15×10^{-4} cm. The absorption of light in this region is associated with changes in vibration and rotation of the molecule concerned. Pure rotation spectra can be observed in the far-infrared region but this is of less value in determining the structure of organic molecules and its observation involves formidable experimental difficulties.

The bond joining two atoms will resist extension and compression. We can compare a simple diatomic molecule with two masses connected by a spring. If the spring is stretched and then released, the masses execute simple harmonic motion about their equilibrium position. The situation in the diatomic molecule is similar and vibration can only occur at certain fixed frequencies. The vibration energy levels are given by $Ev = h(v + \frac{1}{2})$, where $v = 0, 1, 2, \ldots$ (the vibrational quantum number). Notice that this means that even for the lowest vibrational level ($v = 0$) the molecule still possesses vibrational energy, known as the zero point energy. Polyatomic molecules have many modes of vibration. In the water molecule, for example, there are three:

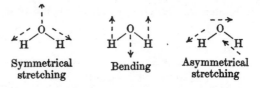

Symmetrical stretching Bending Asymmetrical stretching

To absorb near-infrared radiation a molecule must be able to vibrate in such a way that there is a displacement of electric charge about the central position, i.e. the dipole must change during the vibration. This means that homonuclear diatomic molecules such as H_2, N_2, O_2, etc., are inactive in the infrared. On the other hand, bonds like C—H, N—H, O—H and the carbonyl C=O have a dipole moment which changes when the bonds stretch. The value of infrared spectra for the study of complex organic molecules is greatly enhanced because these complex molecules can be regarded as a heavy mass to which light masses are attached. The spectrum shown in Figure 2.2 is that of the

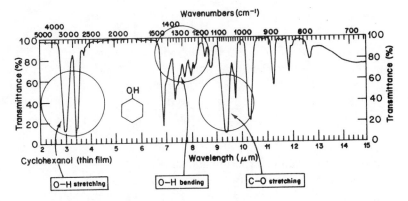

Figure 2.2 Spectrum of cyclohexanol. (*Reproduced by permission of Allyn & Bacon Inc from* IR Spectroscopy, *p. 124.*)

simple alcohol cyclohexanol. The feature to notice is that three of the strong bands can be attributed to particular vibrations in the parent molecule, i.e. the O—H stretching frequency, the O—H bending frequency and the C—O stretching frequency. We shall see later that the other strong absorption bands in the spectrum can be attributed to carbon–hydrogen stretching and bending frequencies.

The great value of infrared spectroscopy is that, depending on their masses, different bonds absorb in different regions of the spectrum (see the Appendix, Table A.1, for a chart of characteristic infrared absorptions).

Nuclear Magnetic Resonance (Proton Magnetic Resonance)

The structure of atomic nuclei is principally the concern of the nuclear physicist. However, some nuclei have a magnetic moment in consequence of their spin. Nuclear spin is governed, like electron spin, by a spin quantum number I. Nuclei which occur frequently in organic compounds are ^{12}C and ^{16}O for which $I = 0$, and ^{1}H, ^{13}C, and ^{19}F for which $I = \frac{1}{2}$.

When such a spinning nucleus is placed in a magnetic field, with the nucleus oriented at an angle θ to the magnetic field, the field will exert a force on the nucleus tending to align the spinning nucleus with the field (Figure 2.3); however, as the nucleus is spinning, the net effect will be to cause the nucleus to precess

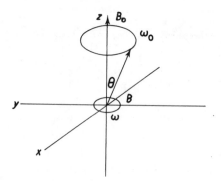

Figure 2.3 Effect of a magnetic field B_0 in a spinning nucleus.

around the axis of the field with an angular velocity ω_0. This precession is analogous to the precession of a spinning gyroscope when it is allowed to topple in the earth's gravitational field.

The angular velocity ω_0 is proportional to the intensity of the applied field B_0; the constant of proportionality is known as the gyromagnetic ratio γ, whence $\omega_0 = \gamma B_0$.

If a small oscillating field B is applied perpendicularly to B_0, this field B will tend to tilt the spinning nucleus towards the xy plane. It will be effective only when the oscillating field B is rotating about the axis of B_0 with the same angular velocity as the precession velocity ω_0. Under these circumstances there will be an exchange of energy between the rotating field B and the precessing nucleus. This represents a familiar resonance phenomenon; hence the name nuclear magnetic resonance (n.m.r.).

The way in which a nucleus aligns itself with the applied magnetic field is governed by its spin. The nucleus can assume any one of $2I + 1$ orientations. For the proton, $I = \frac{1}{2}$, and there are therefore only $2(\frac{1}{2}) + 1 =$ two possible orientations of the proton, corresponding to magnetic quantum numbers of $m = +\frac{1}{2}$ and $m = -\frac{1}{2}$. One of these two orientations can be considered to be parallel to the applied field ($m = +\frac{1}{2}$) and the other antiparallel ($m = -\frac{1}{2}$). These two orientations have slighly different energies (see Figure 2.4). Excitation of the nucleus results in the transition of the nucleus to the energetically less favourable orientation.

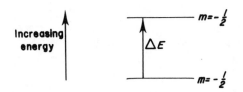

Figure 2.4 Energy levels of a proton in a magnetic field.

The difference in energy ΔE between the two states is proportional to the strength of the field B at the nucleus:

$$\Delta E = \gamma h B / 2\pi \tag{2}$$

As with infrared and ultraviolet spectra, ΔE is associated with a corresponding frequency of electromagnetic radiation:

$$\Delta E = h\nu \tag{3}$$

From equations (2) and (3) we obtain

$$\nu = \gamma B / 2\pi \tag{4}$$

For a hydrogen nucleus in a magnetic field of 1.4 Tesla(T), the energy ΔE corresponds to frequencies in the short-wave radio region (60 Hz).

Experimentally it is easier to vary the applied magnetic field than the additional oscillating field, so that a molecule is irradiated with a fixed frequency ν_i, and the applied magnetic field B_0 is varied until $\nu_f = \gamma B_0 / 2\pi$. At this point the sample absorbs energy. Any further change in B_0 will result in a decrease in absorption. An absorption spectrum can be presented as a plot of absorption against B_0 (Figure 2.5).

Absorption of energy will be observed if there is an excess of nuclei in the orientation of lower energy. The distribution of nuclei between the two levels is governed by the normal (Maxwell–Boltzmann) law: $n_1/n_2 = \exp(-\Delta E/RT)$. The value of B_0 is made as high as possible in order to give a significant difference in population between the higher and the lower energy levels.

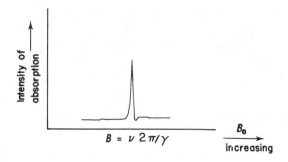

Figure 2.5 Nuclear magnetic resonance spectrum of a single nucleus.

So far we have neglected the extranuclear electrons. The electrons that surround a hydrogen atom and bind it to the rest of the molecule shield the nucleus, so that the effective field experienced by the nucleus is slightly less than the applied field. The screening effect of the electrons arises from interaction between the magnetic field of the electrons and the applied magnetic field.

The field experienced by a particular proton will therefore depend upon the electron density around the proton, and this density will depend on the atom to which the proton is attached. The resonance signals produced for differently bound hydrogen atoms occur at different field strengths. These differences in field strengths at which signals are obtained for protons in different molecular environments are called chemical shifts. The nuclear magnetic resonance spectrum of diacetone alcohol is shown in Figure 2.6. The proton of the hydroxyl group is the least shielded of the four types of protons in the molecule (the electronegative oxygen atom withdraws electrons from the neighbourhood of the hydrogen atom of the hydroxyl group). Next appears the CH_2 group, not so strongly deshielded as the OH proton but, nevertheless, being flanked by an electron-withdrawing C=O group on one side and the C—OH group on the other, more deshielded than the methyl group attached to the C=O group. The absorption remaining is due to the protons of the pair of methyl groups at the end of the chain.

Chemical shifts are measured relative to a standard, usually the absorption of the protons in tetramethylsilane. These protons

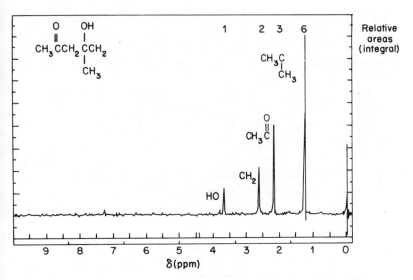

Figure 2.6 Nuclear magnetic resonance spectrum of 4-hydroxy-4-methylpentan-2-one (diacetone alcohol).

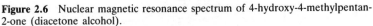

absorb at a higher frequency than the protons of the majority of organic compounds. The chemical shift δ is expressed as

$$\delta = \frac{(\nu_{sample} - \nu_{reference})}{\nu_{oscillator}} \times 10^6 \text{ Hz ppm}$$

The intensity of each resonance line is proportional to the number of corresponding protons in the sample, and it is usually recorded on the spectrum as a line (integral) representing the area under each peak.

As for ultraviolet and infrared spectra, a list of values can be assigned to characteristic structures—in this case, different structural environments for a proton. From the position of an absorption line in an n.m.r. spectrum, a guess can be made as to whether the proton is in a strongly electron-deficient environment (e.g. the acidic hydrogen of a carboxylic acid) or not (e.g. the methyl group in ethanol). Aromatic protons appear at low δ values owing to the ring current which results from the action of the applied magnetic field on the closed loops of π electrons above and below the plane of the aromatic ring (cf. Figure 2.7). This interaction produces a magnetic field which adds to the applied magnetic field experienced by the protons attached to the aromatic ring.

The values given in Table A.2 in the Appendix are only approximate since individual structures affect the environment of a particular proton in different ways, but the table can be used in conjunction with the relative peak areas to detect the presence of characteristic groups; for example, O—CH$_3$ protons usually appear between $\delta = 4.0$ and 3.6 whereas the protons of the CH$_3$ group in saturated chains appear around $\delta = 1.0$ to 0.8 ppm.

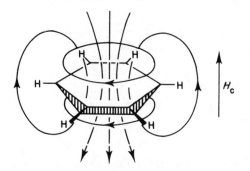

Figure 2.7 Deshielding of aromatic protons by the ring current effect.

Spin–spin Coupling

Nuclear magnetic resonance spectra of organic molecules are not usually as simple as illustrated in Figure 2.6. For example, the spectra of acetaldehyde and of ethyl iodide are shown in Figures 2.8 and 2.9.

From the integrals and the chemical shifts, the absorptions can be assigned, but the question remains, why should the lines be split? To answer this question we must return to the original reason for the occurrence of n.m.r. lines, namely the fact that protons possess a magnetic moment due to nuclear spin (I for protons $= \frac{1}{2}$). This nuclear magnetic moment exerts an influence over its immediate surroundings, and thus the three protons of the CH$_3$ group in acetaldehyde are under the influence of a small magnetic field due to the spinning nucleus of the CH=O proton.

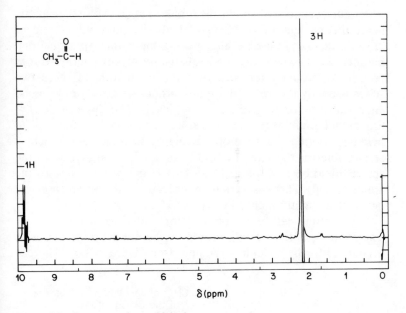

Figure 2.8 Spectrum of acetaldehyde.

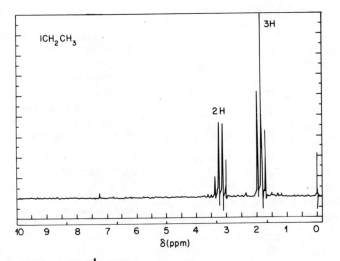

Figure 2.9 Spectrum of iodoethane (ethyl iodide).

When the acetaldehyde molecule is placed in an external magnetic field, the nucleus of the CH=O hydrogen may adopt one of $2I + 1$, i.e. two, possible alignments, one with a spin quantum number $m = +\frac{1}{2}$ and the other of slightly higher energy with $m = -\frac{1}{2}$. At ordinary temperatures, the population of these two states is nearly the same (not quite, otherwise there would be no energy absorption to give an n.m.r. signal). The three protons of the methyl group experience either a small addition to the applied field because the CH=O proton is aligned with the applied field— responsible for the line at $\delta=2.17$—or a slight diminution of the applied field because the field of the CH=O proton opposes the applied field. The absorption of energy in the latter case will occur at a slightly higher field ($\delta=2.14$).

The splitting can be pictorially represented as in Figure 2.10. The splitting of the CH=O proton by the methyl protons can be accounted for by a similar argument (see Figure 2.10b). There is

2.15δ

J = 2.84 Hz

(a) Splitting of the CH_3 proton in acetaldehyde

CH$_3$ absorption in the absence of the CHO proton.

CH$_3$ absorption in the two possible spin orientations of the CHO proton.

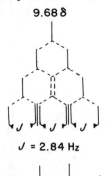

9.68 δ

J = 2.84 Hz

(b) Splitting of the CHO proton in acetaldehyde

CHO absorption in the absence of the CH$_3$ protons.

CHO absorption in the presence of one of the CH$_3$ protons.

CHO absorption in the presence of two of the CH$_3$ protons.

CHO absorption in the presence of all of the CH$_3$ protons.

Ratio of line intensities
1:3:3:1

In general, this ratio for an n line multiplet is given by the coefficients of the series $(x + 1)^n$

Figure 2.10 Diagram illustrating splitting of proton signals.

no splitting between the three protons of the methyl group, since all protons are in equivalent environments (remember that at ordinary temperature there is complete freedom of rotation about a carbon–carbon single bond).

Splitting is most frequently encountered between protons on adjacent atoms, and there is virtually no interaction between hydrogens that are located on carbon atoms further apart unless there is a multiple bond between two of these carbon atoms.

With ethyl iodide the spectrum is more complex in that the CH_3 is adjacent to two protons, each of which may be aligned either with or against the applied field. The interaction may be represented as in Figure 2.11. The protons of the CH_3 group interact with the two protons of the CH_2 group and appear as a triplet (relative intensities 1 : 2 : 1), with an integral equivalent to three protons.

In general, a proton in the field of n other equivalent protons will be split into $2nI + 1$ lines. For the proton, $I = \frac{1}{2}$, so that this expression simplifies to $n + 1$; for other nuclei, for example ^{14}N, where $I = 1$, the splitting is more complex.

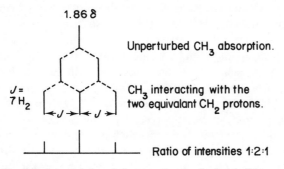

1.86 δ

Unperturbed CH_3 absorption.

$J = 7\,H_2$

CH_3 interacting with the two equivalant CH_2 protons.

Ratio of intensities 1:2:1

Figure 2.11 Splitting of the methyl group signals of ethyl iodide.

Coupling Constants

The magnitude of the splitting described above depends on the intensity of the interaction between protons, and, in contrast to the chemical shift, is independent of the applied field. The magnitude of the splitting is called the coupling constant, J. J values are expressed in cycles per second or hertz (Hz).

In the spectrum of acetaldehyde, the splitting of the aldehyde proton by the methyl protons, i.e. the coupling constant $J(H,$

CH_3) (~3 Hz), is the same as the coupling constant $J(CH_3, H)$ for the methyl absorption (~3 Hz) since the interaction of one type of proton with another is transmitted equally either way.

The coupling of two protons in a similar environment is more complex and will not be dealt with here.

Chemical Exchange

Labile protons, such as the hydroxyl proton of alcohols and phenols, which tend to exchange with neighbouring molecules in the presence of catalysts, e.g.:

$$C_2H_5O{-}H \rightleftharpoons C_2H_5O + H^+$$
$$H^+ \qquad H$$

have chemical shift values that vary according to the purity of the sample and are concentration-dependent. There will be no spin–spin splitting between such protons and neighbouring protons, provided that the rate of chemical exchange is faster than the coupling constant between these two types of protons. Traces of catalysts, H^+ or OH^-, are sufficient to initiate this exchange so that only in very pure samples can coupling between an OH proton and neighbouring C—H protons be observed. Rapid proton exchange is also observed with the proton bound to nitrogen of amines. These functional groups exchange protons with ^2H (deuterium) in the presence of ^2H_2O. Thus the OH or NH absorptions can be suppressed by the addition of ^2H_2O to the solution the ^1H spectrum of which is being measured. ^2H has a resultant nuclear spin $I = 1$ but the absorption is in a different region of the spectrum and therefore does not resonate at frequencies appropriate for ^1H spectra.

^{13}Carbon Nuclear Magnetic Resonance

The nucleus ^{13}C has a spin quantum number $I = \frac{1}{2}$ similar to ^1H which also has $I = \frac{1}{2}$. However, to observe a ^{13}C resonance spectrum of an organic compound a different frequency is needed; at 1.4 tesla (T), ^{13}C resonates around 15 MHz compared to ^1H which resonates at 60 MHz at the same value of the magnetic field.

At first sight, the observation of a carbon nuclear magnetic resonance spectrum should be unlikely. The ^{12}C nucleus has no resultant spin quantum number, $I = 0$, and therefore is inactive, and the ^{13}C nucleus which has $I = \frac{1}{2}$ similar to ^1H is present to the extent of just 1 per cent in any ordinary organic compound. However, with sensitive spectrometers, capable of accumulating many spectra and thus distinguishing weak absorptions from the background electronic noise, this low abundance of the magnetically active isotope is an advantage. The chance of finding two ^{13}C nuclei adjacent to one another is 1 per cent of 1 per cent so that there is no splitting caused by adjacent ^{13}C nuclei.

The position of absorption of a ^{13}C nucleus is affected by the electronic surroundings of that nucleus: ^{13}C nuclei attached to electron-withdrawing groups experience a greater chemical shift than, for example, ^{13}C nuclei in alkanes. There is a correlation chart (see the Appendix, Table A.3) similar to the chart for chemical shift in ^1H n.m.r. The chemical shift range for ^{13}C in organic compounds is, however, much greater, (0–230 ppm relative to tetramethyl silane as an internal standard). As an example compare the ^1H and ^{13}C spectra of methyl methanoate, shown in Figure 2.12.

In general alkyl carbons absorb in the δ 0–100 ppm range, alkene carbons in the δ 100–210 ppm range and carbonyl carbon nuclei in the δ 170–210 ppm range.

Integration of ^{13}C resonance signals is a less reliable measure of the number of carbons in the same environment than in ^1H magnetic resonance due to the ease of saturation of the low energy level population, and unless special instrumental techniques are used, integrals cannot be used to aid structure determination. In many compounds each carbon is in a unique environment and so the number of absorptions in the spectrum is an indication of the number of carbon atoms of different types present in the molecule.

Coupling between adjacent carbon atoms does not occur, but coupling between the ^{13}C carbon and the protons attached to that carbon atom will occur. This gives rise to a complex spectrum and to simplify interpretation usually all the protons in the spectrum are decoupled by applying a broad frequency irradiation ('noise decoupling') so that all the ^{13}C absorptions appear as singlets. This simplified spectrum when interpreted in conjunction with the ^1H spectrum of the same compound enables proposals for the structure of the compound being examined to be made.

Mass Spectra

The principles underlying the production of a mass spectrum are unlike those underlying the spectra discussed so far, in that the former do not involve absorption of electromagnetic radiation. In a mass spectrometer, a compound is bombarded with electrons, which ionize and fragment the compound. The positively charged fragments are then separated and their relative abundances measured. Thus a mass spectrum of a compound is a chart showing the relative abundance of the various positive ions produced when the compound is bombarded with electrons, these ions being arranged in order of their masses.

The compound is introduced into an evacuated ionization chamber where it is bombarded with a stream of electrons:

$$e^- + XY \rightarrow XY^+ + 2e^-$$
$$XY^+ \rightarrow X^+ + Y$$
$$\text{Fragment}$$
$$\text{ion}$$

These positive ions are drawn out of the ionization chamber by an electric field V that accelerates them down the tube of the mass spectrometer. They thus acquire velocities v that depend on their masses m_1, m_2, m_3, etc.:

$$eV = \tfrac{1}{2}m_1v_1^2 = \tfrac{1}{2}m_2v_2^2 = \tfrac{1}{2}m_3v_3^2 = \text{etc.} \tag{5}$$

These ions are then passed through a magnetic field where they are subject to a force eBv (where B is the magnetic field strength, e the electronic charge and v the velocity of a particular ion). The magnetic field deflects the positive ions along a circular path of radius r and the deflecting force is balanced by the centrifugal force mv^2/r; i.e.

$$eBv = mv^2/r \tag{6}$$

By combining equations (5) and (6) we get

$$m/e = B^2r^2/2V$$

Thus, for fixed values of the field strength B and of the accelerating potential V, ions having different mass/charge ratios ($m/$

e) follow different paths; for a given value of *B* and *V*, only a selected beam of ions will reach the detector slit. The mass spectrum of a sample is scanned by varying either *H* or *B*.

In double-focusing instruments, the resolving power of the spectrometer is increased by passing the beam of positive ions through an electrostatic field to deflect the ions, as well as through a magnetic deflector.

The mass spectrum of an organic compound is presented in the form of a plot of relative ion abundance against the ratio of mass to charge (Figure 2.13). The spectra shown consist of a large number of sharp peaks that vary in *m/e* ratio from the parent molecular ion formed by the initial ionization reaction:

$$XY + e^- \rightarrow XY^+ + 2e^-$$

through various ions formed by fragmentation of the parent ion:

$$XY^+ \rightarrow X^+ + Y$$

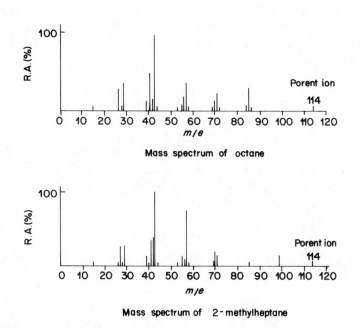

Mass spectrum of octane

Mass spectrum of 2-methylheptane

Figure 2.13 Mass spectra.

These fragment ions may break down further to give smaller fragments. In most cases the parent ion can be seen, although as with 2-methylheptane (Figure 2.13) the signal is often small. From the parent molecular ion peak, the molecular weight of the compound may be determined with an accuracy that enables a distinction to be made between formulae such as $C_{28}H_{50}$(C, 87.0%; H, 13.0%), and $C_{29}H_{52}$ (C, 86.9%; H, 13.1%), a distinction that it would be difficult to make with certainty if combustion analysis alone were used to determine the molecular formula.

Ambiguities arising from different fragments of the same m/e ratio, for example C_2H_5O and C_2H_7N (both give peaks at $m/e = 45$), can be resolved by considering the relative abundances of adjacent minor peaks due to the presence of isotopes of C, H, N and O in the naturally occurring elements. Carbon contains about 1% of ^{13}C, nitrogen 0.4% of ^{15}N, oxygen 0.2% of ^{18}O and hydrogen 0.02% of ^2H (D), so that most organic ion peaks show weak adjacent peaks, one or two units higher in mass, due to the presence of these isotopes. Thus a peak at $m/e = 14$ due to the $^{12}CH_2^+$ ion is accompanied by peaks at $m/e = 15$ ($^{13}CH_2^+$ and $^{12}CHD^+$) and 16 ($^{13}CHD^+$ and $^{12}CD_2^+$), as well as 17 ($^{13}CD_2^+$). The relative ratio of the first three peaks (100 : 1.11 : 0.004) can be calculated from the relative abundances of the isotopes. A peak that did not show adjacent peaks in these ratios would be due to some other molecular ion, or perhaps to a doubly charged ion of twice the mass. In our original example, the relative abundances for C_2H_5O at $m/e = 45$, 46 and 47 are 100 : 2.280 : 0.215, whereas for C_2H_7N they are 100 : 2.665 : 0.0228.

Compounds containing chlorine and bromine can be recognized by clusters of peaks at the molecular ion, or in fragments such as Cl_2^+, Br_2^+, HCl^+ and HBr^+. The element bromine consists of a 1 : 1 mixture of the isotopes ^{79}Br and ^{81}Br so that the clusters comprise a pair of peaks in the abundance ratio of 1 : 1 and separated by $m/e = 2$. Chlorine consists of a 3 : 1 mixture of the isotopes ^{35}Cl and ^{37}Cl so that clusters of peaks in the abundance ratio of 3 : 1, separated by two m/e units, identify the presence of this element.

Molecular formulae can be determined with more reliability than described above by using a high-resolution instrument. The principal peak due to the $C_2H_5O^+$ ion actually has $m/e = 45.034$ ($^1H = 1.00782$; $^{12}C = 12.0000$; $^{16}O = 15.9949$), whereas the principal peak due to the $C_2H_7N^+$ ion actually has $m/e = 45.058$ ($N = 14.0031$). In a double-focusing instrument masses can be determined with an accuracy of 1 in 10,000, so that these two peaks can be distinguished.

In addition to the molecular weight and molecular formula of an organic compound, mass spectra can give an indication of the structure of a molecule from the way in which it breaks down in forming the mass spectrum (the cracking pattern). Structure determination by interpretation of the cracking pattern depends largely on empirical correlations gained by studying the cracking pattern of similar groups of compounds, and usually is not much used for this purpose. In general, the molecular ion mass M^+, or the absence of a molecular ion if the compound is unstable, is all that is needed from a mass spectrum. Infrared spectroscopy and particularly nuclear magnetic resonance spectroscopy provide the majority of the structural information needed for identification of a molecular structure.

Identification and Structural Determination

No understanding of the reactions of organic molecules is possible if the structure of the molecules is unknown. Originally the structure was inferred from chemical reactivity but now structure is determined by physical experiments. Thus at the beginning of each chapter in which a new class of molecules is being discussed, a brief account of the relevant *infrared* and *nuclear magnetic resonance* spectra will be given. *Ultraviolet* (u.v.) spectra are far less useful but the application of ultraviolet spectra to the study of conjugated alkenes is given in Chapter 14.

CHAPTER 3 ———————————————

The Carbon–Hydrogen Bond

As we would expect the spectroscopic characteristics of unsubstituted alkanes are relatively simple. The infrared spectra of the alkanes (see Figure 3.1) exhibit absorption bands which can be attributed to carbon–hydrogen bond stretch (2944–2865 cm^{-1}), carbon–hydrogen bond deformation ($\sim$1460 cm^{-1} and $\sim$1378 cm^{-1}) and absorptions attributable to carbon–carbon single bonds; these latter are too weak to be of much value in structural assignments.

The ^{1}H n.m.r. spectrum of hexane (Figure 3.2) consists of two bands, one attributable to the hydrogen atoms attached to the terminal methyl groups ($\delta \sim 0.90$ ppm) and the other at lower field attributable to the hydrogen atoms bound to the central CH$_2$ groups (δ 1.25 ppm). The band attributable to the terminal methyl groups is a poorly resolved triplet due to spin–spin coupling between the hydrogen atoms on the methyl and the hydrogen

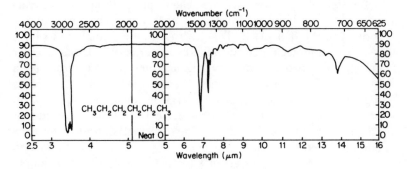

Figure 3.1 Infrared spectrum of hexane (pure liquid).

31

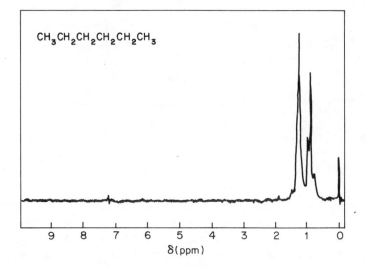

CH₃CH₂CH₂CH₂CH₂CH₃

Figure 3.2 The ¹H n.m.r. spectrum of hexane.

atoms on the adjacent CH_2 group. The band attributable to the CH_2 group is not resolved into fine structure because there are too many very similar interactions, e.g. the hydrogen atoms attached to the second carbon atom are coupled with the hydrogen atoms on the terminal CH_3 group and with the hydrogen atoms on the third carbon atom.

The coupled ^{13}C n.m.r. spectrum of a molecule like heptane is too complicated for most purposes, but if the hydrogen atoms are decoupled (see Chapter 2) we obtain a spectrum in which each type of carbon atom is clearly resolved (Figure 3.3).

Before we can consider any chemical reactions of the carbon–hydrogen bond we must remember why a chemical bond is formed in the first place. Two hydrogen atoms combine to form a hydrogen molecule because the hydrogen molecule has a lower energy than two separate hydrogen atoms (or a proton and a hydride anion). This means that when two hydrogen atoms come together to form a molecule, energy is released. Similarly, in order to break the hydrogen molecule into its two constituent atoms we must supply energy. This energy required to break a bond is called the *bond-dissociation energy*. It is normally written $D(H—H)$ and is expressed in kilojoules per mole, for example $D(H—H) = 432$ kJ mol^{-1}. In methane the carbon–hydrogen bond has a dissociation energy of 427 kJ mol^{-1} [$D(CH_3—H) = 427$ kJ

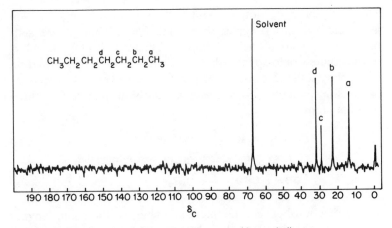

Figure 3.3 The ^{13}C n.m.r. spectrum of heptane (decoupled).

mol^{-1}]. (Notice that this is the bond-dissociation energy for a single–carbon hydrogen bond in methane; it is not the bond-dissociation energy of a carbon–hydrogen bond in a methyl radical, which is slightly less. The average energy required to strip all four hydrogen atoms from carbon is called the bond energy; it is not the same as the bond-dissociation energy, which refers specifically to one carbon–hydrogen bond in methane.)

If we take a mixture of methane and chlorine in the gas phase at room temperature in the dark, no reaction occurs. In the presence of light, however, a rapid reaction ensues and with an excess of methane the main products are methyl chloride and hydrogen chloride, the excess of methane remaining unchanged. The function of the light is to dissociate the chlorine molecule into chlorine atoms. The chlorine molecule absorbs visible light at the violet end of the visible spectrum and near-ultraviolet light. The energy of light in this region of the spectrum is far greater than that necessary to dissociate molecular chlorine, the bond-dissociation energy of which is approximately 281 kJ mol^{-1}. The sequence of reactions that then occurs is shown in the following set of equations:

$$Cl_2 + h\nu \xrightarrow{\ 1\ } 2\ Cl\cdot \qquad \textbf{Initiation}$$

$$Cl\cdot + CH_4 \xrightarrow{\ 2\ } CH_3\cdot + HCl \left.\vphantom{\begin{array}{c}1\\1\end{array}}\right\} \textbf{Propagation} \qquad \Delta H = -8\ \text{kJ mol}^{-1}$$

$$CH_3\cdot + Cl_2 \xrightarrow{\ 3\ } CH_3Cl + Cl\cdot \qquad\qquad\qquad \Delta H = -96\ \text{kJ mol}^{-1}$$

The first feature to notice about these reactions is that as soon as reaction 1 has occurred once, reactions 2 and 3 are self-propagating, so that one quantum of light absorbed by one chlorine molecule would apparently be sufficient to initiate chlorination of all the methane present. In practice, however, this sequence of reactions, called a *chain reaction*, can be terminated by three possible steps:

$$\text{Cl}^{\cdot} + \text{Cl}^{\cdot} + \text{M} \xrightarrow{\ 4\ } \text{Cl}_2 + \text{M}^* \left.\begin{array}{l} \\ \\ \\ \end{array}\right\}$$

$$\text{CH}_3^{\cdot} + \text{Cl}^{\cdot} \xrightarrow{\ 5\ } \text{CH}_3\text{Cl} \qquad \right\} \text{ Termination}$$

$$\text{CH}_3^{\cdot} + \text{CH}_3^{\cdot} \xrightarrow{\ 6\ } \text{C}_2\text{H}_6$$

In the reaction between methane and chlorine the chain length† can be as long as 10^6, although most reaction chains are much shorter. At present we are most concerned with the thermochemistry of the reactions. The enthalpy of reaction, ΔH, for reaction 2 is -8 kJ mol^{-1} and the enthalpy of reaction for reaction 3 is -96 kJ mol^{-1}. Notice that both these enthalpies of reaction are negative, i.e. the reaction is exothermic, heat being given out during the reaction. This means that as regards energy the reaction is going downhill; i.e. it is a favourable process. Now let us compare chlorination with bromination:

$$\text{Br}_2 + h\nu \xrightarrow{\ 1\ } 2\,\text{Br}^{\cdot}$$

$$\text{Br}^{\cdot} + \text{CH}_4 \xrightarrow{\ 2\ } \text{CH}_3^{\cdot} + \text{HBr} \qquad \Delta H = +54 \text{ kJ mol}^{-1}$$

$$\text{CH}_3^{\cdot} + \text{Br}_2 \xrightarrow{\ 3\ } \text{CH}_3\text{Br} + \text{Br}^{\cdot} \qquad \Delta H = -109 \text{ kJ mol}^{-1}$$

---

* Two chlorine atoms coming together do not stay bound unless they can lose the excess 281 kJ. They can lose this energy by collision with another molecule M (which could be Cl_2, CH_4 or CHCl_3, etc.). Two methyl radicals do not require a 'third body', as M is called, because the energy can be partly dispersed in internal vibrations.

† In the last chapter we used the term *chain length* to refer to the number of carbon atoms joined together in a linear hydrocarbon; e.g. in butane, $\text{CH}_3\text{CH}_2\text{CH}_2\text{CH}_3$, the carbon chain is four atoms long. Unfortunately, the same term is used in reaction kinetics to describe the number of times a repetitive cycle of reactions occurs. Thus reaction 2 followed by reaction 3 is considered to be a unit of the chain reaction, and 'chain length' in this context refers to the number of times reactions 2 and 3 are repeated before the chain is terminated by reactions 4, 5 or 6.

(Only the chain-propagating steps are shown here.) It is plain that reaction 2 in bromination is an *endothermic* process, i.e. heat is absorbed during the reaction. On the other hand, reaction 3 is exothermic, as it was in chlorination. The overall process of converting one molecule of methane and one molecule of bromine into one molecule of methyl bromide and one molecule of hydrogen bromide is exothermic. Experimentally, we find that if a mixture of methane and bromine are illuminated at room temperature, reaction takes place, but only very slowly. This reflects the fact that reaction 2 is an endothermic reaction, i.e. this part of that reaction is energetically an uphill process. In order that bromination of methane may occur at a reasonable rate, it is necessary to heat the reactants to provide the necessary energy.

Fluorination, on the other hand, is extremely rapid, even at low temperatures; the reason for this becomes apparent from the following equations:

$$F_2 + h\nu \xrightarrow{1} 2\,F\cdot$$

$$F\cdot + CH_4 \xrightarrow{2} CH_3\cdot + HF \quad \Delta H = -151 \text{ kJ mol}^{-1}$$

$$CH_3\cdot + F_2 \xrightarrow{3} CH_3F + F\cdot \quad \Delta H = -281 \text{ kJ mol}^{-1}$$

Here, both propagating steps are highly exothermic and the conversion of one molecule of methane into one molecule of methyl fluoride is accompanied by the liberation of 432 kJ. This massive release of energy together with the fact that the bond-dissociation energy of fluorine is only 155 kJ mol^{-1} results in *thermal branching* (meaning that additional molecules of fluorine are dissociated thermally, so starting off new chains). It is easy to see that this thermal branching is a self-increasing process and ultimately fluorination results in an explosion unless a great excess of some inert gas such as nitrogen is also present to absorb the heat evolved.

In all these halogenation processes reaction 2 is the rate-determining step. We have argued that bromination proceeds only very slowly at room temperature because reaction 2 is endothermic. However, although chlorination is rapid at room temperature it is accelerated by raising the temperature; even though fluorination is so rapid at room temperature, it becomes faster still if the temperature is raised. This suggests that although the reaction is exothermic it still requires a supply of energy before it will take place. We can illustrate this by the following scheme which represents three stages in reaction 2

$$
\underset{H}{\overset{H}{\underset{\diagdown}{\diagdown}}}{C}\!-\!H + Cl\cdot \longrightarrow \underset{H}{\overset{H}{\underset{\diagdown}{\diagdown}}}{C}\cdots H\cdots Cl \longrightarrow \underset{H}{\overset{H}{\mid}}{C}\cdot + HCl
$$

First we have a methane molecule and a chlorine atom; next a hydrogen atom half-bonded to a methyl radical and half-bonded to a chlorine atom; and finally a methyl radical and a molecule of hydrogen chloride. As the chlorine atom approaches the methane molecule there will be electronic repulsion which will increase as the chlorine atom approaches closer. Finally, the stage is reached when the electrons forming the carbon–hydrogen bond are uncoupled and the new bond is formed between the hydrogen atom and the chlorine atom. Let us consider the reverse process, i.e. the interaction between a methyl radical and a hydrogen chloride molecule. Here, again, as the methyl radical approaches the hydrogen chloride molecule it will be repelled, the repulsion increasing as the species come closer together. Thus, the intermediate species in the three-step reaction sequence above represents a state of high energy. This can be depicted by plotting the energy of the system against the reaction coordinate, the latter representing the progress of the reaction between one molecule of methane and one chlorine atom (Figure 3.4). The difference in energy between the initial state and the final state is equal to ΔH, the enthalpy of reaction. The difference in energy between the initial state and the transition state is equal to E, the activation energy. This is the energy that must be supplied before the reaction occurs. Figures 3.5 and 3.6 represent, respectively, the potential energy situation in reaction 2 for bromination (an endothermic reaction) and in reaction 2 for fluorination (a highly exothermic reaction). Notice that in bromination ΔH is positive and the activation energy E is very large. In fluorination, on the other hand, ΔH is large and negative and the activation energy is very small. Reactions that involve only the making of bonds, such as the recombination of chlorine atoms (reaction 4 in the chlorination chain mechanism) have no activation energy and the rate of this reaction is independent of the temperature. All reactions which involve the breaking as well as the making of bonds have an activation energy, even though they are extremely exo-

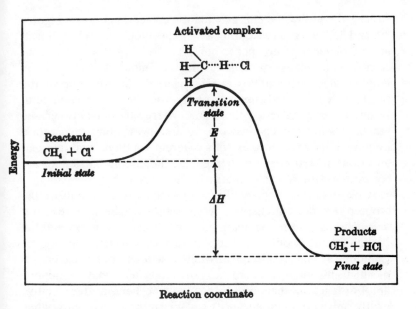

Figure 3.4

thermic as in the case of fluorination, and these reactions will all be accelerated by raising the temperature, no matter how fast they may be.

The rate of a chemical reaction depends not only upon the electronic energy situation but also on the probability that the reactants will collide and, what is more, collide with an orientation relative to one another than enables them to react. When dis-

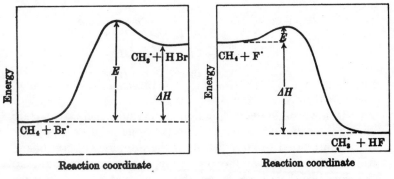

Figure 3.5   Figure 3.6

cussing the electronic energy situation as represented by Figures 3.5 and 3.6, we are considering a methane molecule and a halogen atom isolated in free space; however ordinary practice is concerned with a large number of methane molecules and a large number of chlorine atoms mixed together, and the rate of the reaction then depends upon the properties of the whole system. In any chemical reaction there is an energy barrier between the reactants and products; because we are always concerned with a large number of molecules with different rotational, vibrational, and translational degrees of freedom, the energy barrier we are concerned with is a free energy barrier.

A discussion on the significance of free energy is outside the scope of the present chapter. At present we will simply regard it as a composite term including btoh the electronic energy and the probability. It is only valid to consider the electronic energy in isolation, as was done above, when reactions occur by identical mechanisms, and even here the probability term may be important. For reactions that occur by different mechanisms their relative exothermicity or endothermicity can give no guide to their relative rates.

So far we have considered the chlorination of methane which can yield, as initial product, only one monochlorinated product, namely chloromethane (methyl chloride). Similarly there is only one product from the monochlorination of ethane, chloroethane (ethyl chloride). However, with propane, two products are possible:

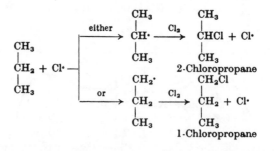

This is a very common situation in organic chemistry. Here, having a reaction from which more than one product is possible, we are interested to know which of the isomeric chloropropanes will be formed. From the discussion concerning enthalpies of reactions, we should expect that the chlorine atom would abstract the most weakly bound hydrogen atom. In practice the reaction yields approximately equal amounts of the two propyl chlorides, but there are only two hydrogen atoms on the central carbon atom whereas there are six identical hydrogen atoms on the two terminal primary carbon atoms. (A CH_3 group is called a primary group, a CH_2 group is called a secondary group and a CH group is called a tertiary group). Generally, in saturated hydrocarbons primary carbon–hydrogen bonds are stronger than secondary bonds, which are stronger than tertiary bonds. Thus a chlorine atom abstracts the more weakly bound secondary hydrogen atoms in propane three times faster than it abstracts primary hydrogen.

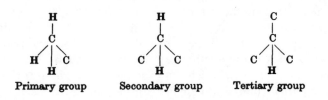

Primary group   Secondary group   Tertiary group

In bromination, reaction 2 is endothermic and the strength of the C—H bond which is being broken will have a much greater effect on the course of the reaction; as a result bromination is very much more selective. In bromination at 150°C the ratio of 2-bromopropane to 1-bromopropane is of the order of 30 : 1 (i.e. the rate of replacement of the hydrogen on the secondary carbon atom is approximately 90 times faster than the rate of replacement of a hydrogen atom on a primary carbon). In fluorination the reaction is highly exothermic and very unselective; slightly more 1-fluoropropane than 2-fluoropropane is obtained but only because the statistical probability of 3 : 1 in favour of primary attack outweighs the actual relative rate of attack, which is only $1\frac{1}{2}$: 1 in favour of the secondary position.

The only reactions discussed above for the aliphatic hydrocarbons are halogenations. The ordinary powerful chemical reagents, such as sulphuric acid, nitric acid, hydrochloric acid, sodium hydroxide, etc., have no effect on alkanes at normal temperatures. This is because they are ionic reagents and the C—H bond does not break ionically at all readily. Consider a molecule AB where the two atoms A and B are bound together by the sharing of a pair of electrons; this can break in three ways as depicted in the following diagram:

$$A:B \longrightarrow A^+ + :B^-$$
$$A:B \longrightarrow A:^- + B^+$$
Heterolysis giving ions

$$A:B \longrightarrow A\cdot + B\cdot$$
Homolysis giving free radicals

Sulphuric acid and sodium hydroxide readily undergo heterolysis giving ions, and the common reactions of these reagents are ionic reactions. It was emphasized in the previous chapter that the carbon–hydrogen bond is non-polar and does not undergo ionic reactions. It does, however, undergo free-radical reactions and the halogenations are examples of this. From what was said in the previous chapter it will also be apparent that the carbon–carbon bond in the alkanes is similarly a non-polar bond and likely to be unaffected by ionic reagents. It is because alkanes are unaffected by the normal ionic reagents that they are often described in textbooks as being chemically inert. In fact this is not really true. The alkanes are not attacked by ionic reagents but they react readily with free radicals and atoms. The normal combusion process by which, for example, petroleum is consumed in an internal combustion engine is principally a free-radical process and the main constituents of petroleum are alkanes. Such hydrocarbons are not inert substances but extremely reactive ones in the appropriate kind of reaction, and this shows how important it is to understand the mechanism of any reactions discussed. Throughout this book we shall consider reactions, how they occur and why.

Nomenclature

A halogen derivative is designated by taking the name of the hydrocarbon from which it is derived and preceding it by prefixes indicating the nature and number of the halogen atoms. Arabic numbers indicate the position(s) of the substituent(s). Three examples are given below:

$CH_3CH_2CH_2CH_2Cl$	1-Chlorobutane
$CF_3CH_2CH_2CH_2CH_3$	1,1,1-Trifluoropentane
$CH_3CHBrCH_2CHClCH_2CH_2F$	2-Bromo-4-chloro-6-fluorohexane

Problems

1. Suggest a mechanism for the iodination of ethane, clearly distinguishing between initiation, propagation and termination

steps. The approximate bond-dissociation energies are: D(H—I), 297 kJ mol^{-1}; D(C—I), 210 kJ mol^{-1}; D(I—I), 151 kJ mol^{-1}. On the basis of this information what would you predict about the feasibility of the iodination of ethane at room temperature?

2. Predict the principal product from the monobromination and monofluorination of butane and 2-methylpropane.

CHAPTER 4

The Carbon–Halogen Bond

The most important feature of the carbon–hydrogen bond is that it is completely non-polar.

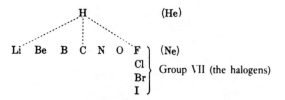

If we look again at the diagram of the first row of the periodic table, and remember that lithium hydride is polarized in the form Li^+H^- while hydrogen fluoride is polarized in the direction H^+F^-, we may expect the carbon–chlorine bond to be polarized in the direction C^+—Cl^-. In the first chapter we emphasized the fact that carbon does not readily form stable anions or cations: the carbon–chloride bond in chloromethane is thus not ionized but only partly polarized as follows:

Infrared spectroscopy is not of great value in determining the structure of simple alkyl halides. Depicted in Figure 4.1 is the infrared spectrum of bromoethane. Notice that the C—H stretch bands ($\sim$2950–2860 cm^{-1}) and the C—H bond deformation ($\sim$1460 cm^{-1} and $\sim$1380 cm^{-1}) are very similar to these bands in hexane. In the halogenoalkanes there is often profusion of

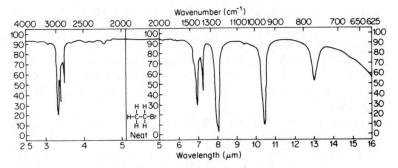

Figure 4.1 Infrared spectrum of bromoethane (pure liquid).

absorption bands throughout the region 1250–770 cm^{-1} and hence little structural information is obtainable.

The ^{1}H n.m.r. spectrum of 1-chloropropane (Figure 4.2) is very characteristic of propyl derivatives. There are three bands which integrate in the ratio 3 : 2 : 2. The band at highest field at approximately δ ~1.0 ppm attributable to the terminal methyl group is a triplet, split by the hydrogen atoms of the central CH$_2$ group. The band at lowest field δ ~3.5 ppm, i.e. the nuclei deshielded by the electronegative chlorine atom, is also a triplet, split by the

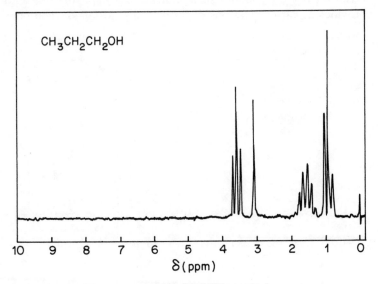

Figure 4.2 The ^{1}H n.m.r. spectrum of 1-chloropropane.

central CH_2 group. The central band, attributable to the central CH_2 group is a multiplet split by both the hydrogen atoms on the methyl group and on the CH_2Cl group.

The decoupled ^{13}C n.m.r. of 2-chlorobutane (Figure 4.3) shows the effect of the relatively electronegative chlorine atom which brings the absorption of the 2-carbon atom substantially further downfield.

In the last chapter the lack of reactivity of alkanes towards ionic reagents such as sulphuric acid and sodium hydroxide was attributed to the fact that carbon–carbon and carbon–hydrogen bonds were non-polar. In *monohaloalkanes* (alkyl halides) we find that the carbon atom attached to halogen carries a small positive charge. Thus an alkyl halide is attacked by species with a negative charge, i.e. by anions. The anion we are most familiar with is the hydroxide anion, OH^-. If a hydroxide ion attacks the carbon atom of chloromethane, then a chloride anion, Cl^-, will be ejected:

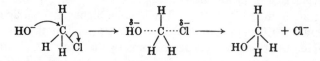

In the diagram the curved arrow from the oxygen atom of the OH ion to the carbon atom is to indicate that the hydroxide anion provides a pair of electrons to form a new carbon–oxygen bond.

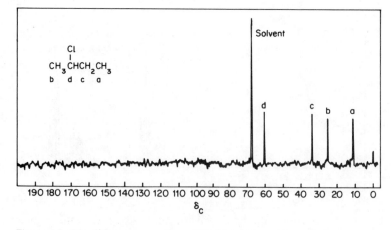

Figure 4.3 The ^{13}C n.m.r. spectrum of 2-chlorobutane (decoupled).

The curved arrow from the bond between carbon and chlorine indicates that the two electrons in the carbon–chlorine bond are both to be donated to the chlorine atom to form a chloride anion. Next we have the activated complex which represents the hydroxide ion coming in on one side of the carbon atom and the chloride anion departing from the other. In the activated complex the three bonds from carbon to hydrogen are approximately planar, at right angles to the plane of the paper. Finally we have the products of the reaction, methyl alcohol (or methanol) together with a chloride anion. This is a general reaction; others can be written in a similar way, such as:

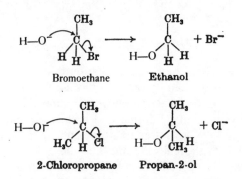

Here again curved arrows are used, so let us define what they represent. A curved arrow represents the *transfer of a pair of electrons* during a reaction process. In the above reactions the hydroxide ion is negatively charged. The oxygen atom of this ion is surrounded by four pairs of electrons, one pair forming the oxygen–hydrogen bond, represented by a single line between the oxygen and the hydrogen, the other three pairs not involved in bonding being shown as pairs of dots:

$$H - \ddot{\underset{\cdot\cdot}{O}}\!:^{-}$$

Such species which supply a pair of electrons to form a new bond are called *electron donors*.

We began by saying that we should expect that the carbon atom attached to a halogen in an alkyl halide would be susceptible to attack by anions, and have shown that when the anion is a hydroxide anion this leads to the formation of alcohols. Table 4.1

Table 4.1 Some common anions.

Acid	Anion	Acid	Anion	Acid	Anion
HCl	Cl^-	HOH	HO^-	HCN	CN^-
HBr	Br^-	CH_3OH	CH_3O^-	HSH	$SH-$
HF	F^-	C_2H_5OH	$C_2H_5O^-$	H_2SO_4	HSO_4^-
HI	I^-	CH_3COOH	CH_3COO^-	$HClO_4$	ClO_4^-

includes some of the common anions. Let us examine the halide anions first. Since the halide anions are similar to one another we should expect the reaction between a fluoride anion and butyl chloride to be a reversible reaction yielding butyl fluoride and a chloride anion:

$$C_4H_9Cl + F^- \rightleftharpoons C_4H_9F + Cl^-$$

Now we know from the previous chapter that the carbon–fluorine bond is stronger than a carbon–chlorine bond and we might therefore expect the reaction between a fluoride anion and butyl chloride forming butyl fluoride and a chloride anion to be exothermic. However, this argument does not take into account the extent to which the chloride and fluoride anions may be bound by the solvent. In practice a fluoride anion in aqueous solution is very tightly bound by water molecules. (Such binding of ions by a solvent is called *solvation*.) This solvation of a fluoride ion is so strong that the fluoride ion becomes extremely unreactive and no reaction can take place between an aqueous solution of sodium fluoride and butyl chloride. However, we should expect the reaction to take place in a solvent which binds (*solvates*) ions less readily, and this is true for dry glycol (glycol has the structure $HOCH_2CH_2OH$). Thus butyl chloride can be converted into butyl fluoride by heating the former with potassium fluoride in dry glycol:

$$C_4H_9Cl + KF \xrightarrow[\text{glycol}]{\text{Dry}} C_4H_9F + KCl$$

Notice that the mechanism of this reaction is the same as that of the replacement of a chlorine atom by a hydroxyl:

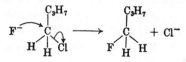

The displacement of chlorine by fluorine is an exothermic process, but that of chlorine by iodine would be endothermic (energetically an uphill reaction):

$$C_5H_{11}Cl + NaI \rightleftharpoons C_5H_{11}I + NaCl$$

Just as the reaction between butyl chloride and a fluoride anion is reversible, so a reaction between pentyl chloride and sodium iodide would be reversible. However on energetic grounds we should expect the equilibrium to lie very much on the side of pentyl chloride and sodium iodide; i.e. the equilibrium constant, K, is much less than 1:

$$K = \frac{[C_5H_{11}I][NaCl]}{[C_5H_{11}Cl][NaI]}$$

As K is a constant, if sodium chloride (NaCl) is removed by some process, then more pentyl iodide must be formed to maintain the equilibrium. Sodium iodide is soluble in acetone, but sodium chloride is not. If we reflux pentyl chloride and sodium iodide in acetone, then, even though at the start the equilibrium is very much in favour of the reactants, as soon as any sodium chloride is formed it will be precipitated, and ultimately the reaction is forced in the direction of pentyl iodide and sodium chloride.

In the third column of Table 4.1, the first acid listed is water and we have already considered the reaction between an alkyl halide and a hydroxide anion. Just as the hydrogen atom in water can formally be replaced by a methyl (CH_3) group to give methanol (CH_3OH), so can the hydrogen atom in the hydroxide anion be replaced by an alkyl radical to give, in the case of methyl, a methoxide anion (CH_3O^-) and of ethyl an ethoxide anion ($C_2H_5O^-$). These anions react in the same fashion as the hydroxyl anion:

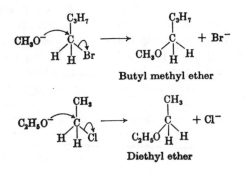

Butyl methyl ether

Diethyl ether

These reactions are quite general.

The hydrogen sulphide anion reacts similarly (the product of this reaction is called a mercaptan, or according to systematic nomenclature, a thiol):

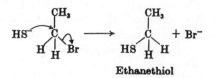

Ethanethiol

The cyanide anion also displaces a halogen in a similar reaction and the product, in this case an alkyl cyanide, is of particular interest because a new carbon–carbon bond is formed:

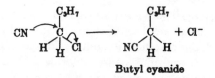

Butyl cyanide

This is an extremely important reaction in the synthesis of organic compounds and can be generalized in the following form:

$$K^{+}CN^{-} \quad R{-}X \xrightarrow[\text{solvent}]{C_{2}H_{5}OH} RCN + K^{+}X^{-}$$

where R represents any alkyl group, potassium cyanide a source of cyanide ions and X any halogen.

The acetate anion also displaces a halogen anion:

$$Na^{+}CH_{3}COO^{-} \quad C_{3}H_{7}{-}I \longrightarrow CH_{3}COOC_{3}H_{7} + Na^{+}I^{-}$$

Propyl ethanoate

This still leaves two anions from Table 4.1 whose displacement reaction has not yet been described. The reason for this is that these are anions of strong acids while the others (except the halogen anions) are anions of weak acids. Perchloric acid is a very strong acid because the perchlorate anion firmly holds the pair of electrons which form the oxygen–hydrogen bond in undissociated acid. The stronger the acid the more firmly the anions hold this electron pair. All the displacement reactions described involve the donation of an electron pair to the slightly electropositive carbon atom attached to the halogen atom in the alkyl halide. The stronger the acid from which the anion is derived, the less readily will that anion donate its electron pair. In general, therefore, it is only the anions of weak acids which will displace halogen anions from the alkyl halides.*

We have been careful to emphasize that it is the electron pair which the incoming anion donates; thus we have depicted the displacement reaction as

$$[H\ddot{\ddot{O}}]^- \overset{\frown}{\underset{Cl}{C}} \longrightarrow H\ddot{\ddot{O}} \overset{|}{C} + [\ddot{\ddot{Cl}}]^-$$

If this is correct, it is the availability of an electron pair which is important rather than the negative charge; thus a compound such as ammonia which has an electron pair readily available for donation would undergo this kind of reaction:

$$H_3N\colon \overset{\frown}{\underset{Cl}{C}} \longrightarrow H_3\overset{+}{N}\overset{|}{\underset{}{C}} + Cl^-$$

Why does water, which has two non-bonded pairs of electrons on the oxygen atom, not displace the halogen anion in the same way? The answer is that ammonia is a stronger base than water which is only a weak one. Nitrogen is less electronegative than oxygen, and the lone pair on the nitrogen atom in ammonia is readily available for bonding with a proton whereas the two pairs of electrons on the more electronegative oxygen are not. Let us now look further into the reaction of ammonia with an alkyl halide:

$$\overset{\frown}{N}H_3 \quad C_2H_5{-}Br \longrightarrow C_2H_5NH_3^+ + Br^-$$

$$C_2H_5NH_3^+ + NH_3 \rightleftharpoons C_2H_5NH_2 + NH_4^+$$

Ethylamine

The first product from the reaction of ammonia with ethyl bromide is ethylammonium bromide, but in the presence of an excess of ammonia we may expect the proton on the ethylammonium ion

---

* This discussion should not be taken to imply anything about the rate of substitution reaction with different ions. The readiness of an anion to donate a pair of electrons is only one of a number of factors (e.g. we have already discussed the strength of the new bond being formed and also the solvation of the anion).

to equilibrate with ammonia molecules to yield free ethylamine and the ammonium cation. Ethylamine has a non-bonded pair of electrons, just as ammonia has, so we may expect further reactions:

$$C_2H_5\ddot{N}H_2 \quad C_2H_5{-}Br \longrightarrow (C_2H_5)_2NH_2{}^+ + Br^-$$

$$(C_2H_5)_2NH_2{}^+ + NH_3 \rightleftharpoons (C_2H_5)_2NH + NH_4{}^+$$

Diethylamine

The product of this reaction, diethylamine, also has a non-bonded pair of electrons and we may expect further reaction:

$$(C_2H_5)_2\ddot{N}H \quad C_2H_5{-}Br \longrightarrow (C_2H_5)_3NH^+ + Br^-$$

$$(C_2H_5)_3NH^+ + NH_3 \rightleftharpoons (C_2H_5)_3N + NH_4{}^+$$

Triethylamine

The product of this further reaction, triethylamine, still has a pair of non-bonded electrons and we must therefore expect yet a further reaction:

$$(C_2H_5)_3\ddot{N} \quad C_2H_5{-}Br \longrightarrow (C_2H_5)_4N^+ \; Br^-$$

Tetraethylammonium bromide

The product of this last reaction, tetraethylammonium bromide, no longer has a pair of non-bonded electrons and the reaction can now proceed no further. The product of the first reaction, ethylamine, is a primary amine. That of the second stage, diethylamine, is a secondary amine and that of the third stage of the reaction, triethylamine, is a tertiary amine, and the product of the final stage, tetraethylammonium bromide, is a quaternary salt. In general the reaction between an alkyl halide and ammonia will lead to a mixture of all four products.

We have now discussed at some length reactions in which a species with a pair of non-bonded electrons attacks the slightly electropositive carbon atom in an alkyl halide and displaces the halogen atom which is ejected as a halide anion:

$$Y: + \; R{-}X \longrightarrow Y \cdots R \cdots X \longrightarrow Y{-}R + X:$$

We began this discussion by saying that carbon did not readily form a carbocation and that, although the carbon–halogen bond was polarized, no free ions were formed. In solvents of high dielectric constant, separation of charge is facilitated and we can expect a greater degree of ionization. Under these circumstances and provided that the anion comes from a very strong acid and that the alkyl group contains substituents to stabilize a positive charge, ionization may occur to a very limited degree. The difficulties in forming a carbocation suggests that it is a high-energy species, i.e. a most reactive entity. Once formed, it will react extremely rapidly with any species able to donate an electron pair. We can thus visualize a two-step process:

$$R\overset{\frown}{-}X \underset{\text{Fast}}{\overset{\text{Slow}}{\rightleftharpoons}} R^+ + X^-$$

$$R^+ + Y^- \xrightarrow{\text{Fast}} R-Y$$

Substitution of an alkyl halide by this mechanism is rare among primary alkyl halides and most common with tertiary alkyl halides. This is because the stability of aliphatic carbocations is in the following order: tertiary more stable than secondary, which is more stable than primary, with a methyl ion least stable. An example of this reaction is hydrolysis of 2-bromo-2-methyl-propane:

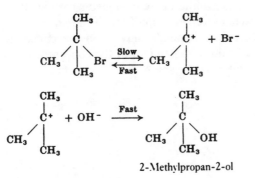

2-Methylpropan-2-ol

The free carbocation has only a *transient existence*. It is most important to appreciate the difference between these carbocations which have a transient existence in a reaction process from the stable ions such as hydroxonium, chloride, etc. Displacement reactions involving a carbocation are much less common than the displacement reactions discussed earlier in this chapter. The former occur when the alkyl halide is tertiary or when the displaced anion, instead of being a halogen, is an anion of a stronger acid such as a sulphonic acid. Further discussion of this subject is outside the scope of the present chapter.

All these reactions have been concerned with displacement of a halogen atom by some species carrying a pair of non-bonded electrons. Such species are called electron-pair donors, or electron donors for short, because they have an electron pair which is available for bonding. These reactions are called 'bimolecular displacement reactions'. We have discussed two mechanisms for this substitution at a carbon atom. In the first, an electron donor with its lone pair of electrons attacks the slightly electropositive carbon of the alkyl halide and the halogen anion is ejected. The rate of such a reaction depends upon the concentration of the electron donor and upon the concentration of the alkyl halide. It is a bimolecular reaction:

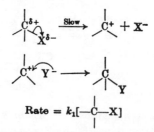

$$\text{Rate} = k_2[\text{Y}^-][\underset{|}{\overset{|}{-\text{C}-}}\text{X}]$$

In the second, less common mechanism the alkyl halide first ionizes to a very limited extent to give a carbocation, which once formed reacts extremely rapidly with the anion. The rate of this reaction therefore depends only upon the concentration of the alkyl halide and is independent of the concentration of the anion. It is a unimolecular reaction:

$$\text{Rate} = k_1[\underset{|}{\overset{|}{-\text{C}-}}\text{X}]$$

The common and more important reaction involving the push–pull transition state is called a bimolecular displacement, while the less common reaction involving a transient carbocation involves only one species in the rate-determining step and is therefore called a unimolecular reaction.

Problems

1. What reaction, if any, would you expect between bromoethane and the following species? Indicate any peculiarities about the conditions necessary for reaction.

(a) OH^- (b) H_3O^+ (c) $Cl\cdot$

(d) CN^- (e) NH_3 (f) HNO_3 (aqueous)

(g) F^- (h) $CH_3CH_2O^-$ (i) I^-

2. Identify the compound responsible for the following mass spectrum:

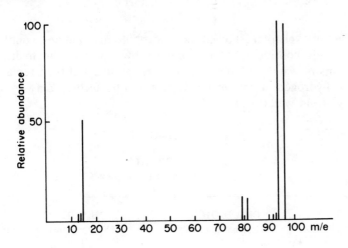

CHAPTER 5 ————————————————————

Alcohols and the OH Group and Amines and the NH₂ Group

The first chapter showed how alcohols and amines could be formally considered to be built up by replacing the hydrogen atoms of water and ammonia by CH_3 groups and then replacing the hydrogen atoms on the CH_3 group by further CH_3 groups; e.g. from water:

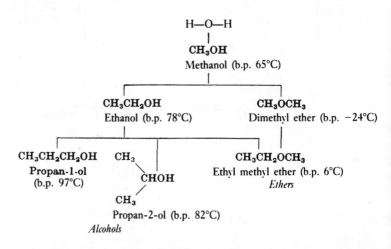

The striking feature of the infrared spectra of butan-1-ol (Figure 5.1) and compounds such as 1-aminobutane (Figure 5.2) containing an hydroxy group attached to a saturated —CH_2 group is a broad band at 3600–3200 cm^{-1} attributable to the stretching frequency of the O—H group. There are corresponding bands in the spectra of primary aliphatic amines at 3400 and 3330 cm^{-1} which are attributable to the asymmetric and symmetric N—H stretching. In the spectra of alcohols there is a band at 1400–1300

NH$_3$
|
CH$_3$NH$_2$
Methylamine (b.p. −6°C)

CH$_3$CH$_2$NH$_2$ (CH$_3$)$_2$NH
Ethylamine (b.p. 10.5°C)   Dimethylamine (b.p. 7°C)

CH$_3$CH$_2$CH$_2$NH$_2$ CH$_3$ CH$_3$CH$_2$ (CH$_3$)$_3$N
Propylamine   Trimethylamine
(b.p. 49°C)   CHNH$_2$ NH   (b.p. 3°C)

           CH$_3$ CH$_3$
2-Aminopropane   N-Methylethylamine
(b.p. 34°C)   (b.p. 36°C)

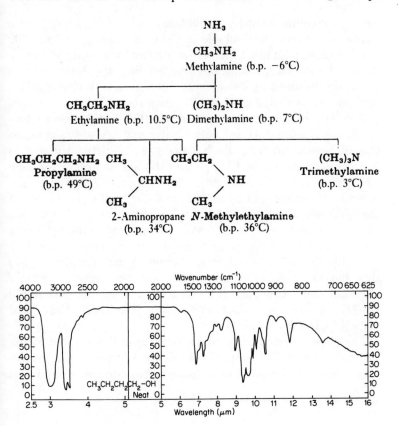

Figure 5.1 Infrared spectrum of butan-1-ol.

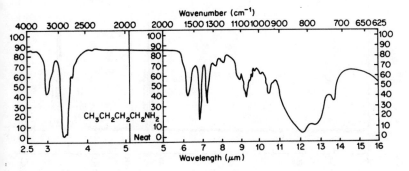

Figure 5.2 Infrared spectrum of 1-aminobutane.

cm^{-1} of medium intensity attributable to O—H bending, and a strong broad band attributable to C—O stretching at 1250–1000 cm^{-1}. In primary amines such as 1-aminobutane (Figure 5.2) there is, in addition to a strong in-plane bending at ~1600 cm^{-1}, a broad diffuse band at 900–650 cm^{-1} which is very characteristic and can be attributed to the H$_2$N group (out-of-plane bending).

The ^{1}H n.m.r. spectra of propan-1-ol (Figure 5.3) and 1-aminopropane (Figure 5.4) are very similar and are also similar to that of 1-chloropropane. Each contain two triplets attributed to hydrogen atoms on the CH$_3$ group (3H) and the other terminal CH$_2$ group (2H). In between is the multiplet (2H) due to the central CH$_2$ group split by the terminal CH$_2$ and CH$_3$ groups. In the spectrum of propan-1-ol there is a singlet at δ 3.1 ppm due to the proton bound to the oxygen atom, and similarly in the spectrum of 1-aminopropane there is a strong singlet at δ 1.05 ppm due to the hydrogen atoms bound to the nitrogen. Notice that the more electronegative oxygen atom causes a greater chemical shift. Compare the chemical shift of the substituted —CH$_2$ group in ClCH$_2$CH$_2$CH$_3$, HOCH$_2$CH$_2$CH$_3$ and H$_2$NCH$_2$CH$_2$CH$_3$.

The effect of relative electronegativity is well illustrated by the ^{13}C n.m.r. spectra of propan-1-ol (Figure 5.5) where the absorption due to the terminal substituted carbon atom is shifted well

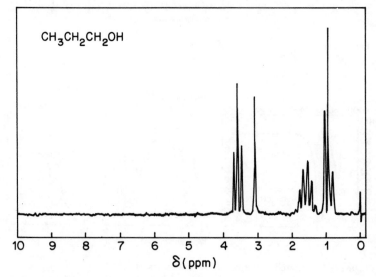

Figure 5.3 ^{1}H n.m.r. spectrum of propan-1-ol.

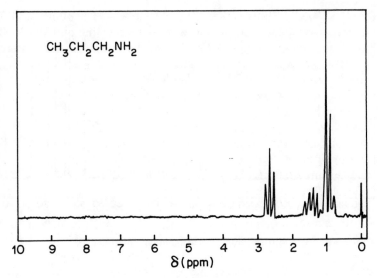

Figure 5.4 ^{1}H n.m.r. spectrum of 1-aminopropane.

downfield, whereas the absorption due to the terminal substituted carbon atom in 1-aminohexane (Figure 5.6) is only slightly downfield from the adjacent CH$_2$ group.

Before considering the chemistry of the OH and NH$_2$ groups attached to a carbon chain, we will briefly discuss the characteristic chemical properties of water and ammonia.

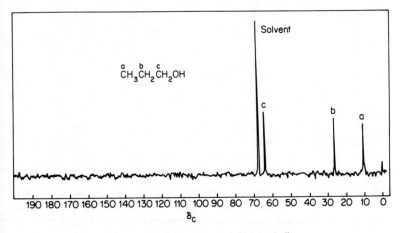

Figure 5.5 ^{13}C n.m.r. spectrum of propan-1-ol (decoupled).

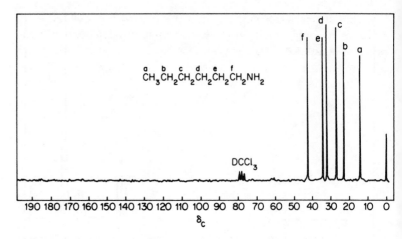

Figure 5.6 ^{13}C n.m.r. spectrum of 1-aminohexane (decoupled).

Among the hydrides of the elements of the first row of the periodic table, starting at the right with group VII, there is hydrogen fluoride, a low-boiling extremely acidic liquid; in group VI, water, boiling at 100°C; in group V, ammonia, a low-boiling extremely basic liquid; and finally in group IV, methane, a very low-boiling gas, inert to ionic reactions. In methane all the eight electrons in the outer shell of the carbon atom are shared with hydrogen. In ammonia there is one pair of non-bonded electrons. In water there are two pairs and in hydrogen fluoride three pairs of non-bonded electrons in the outer shell. The non-bonded electrons in hydrogen fluoride are so tightly held by the fluorine atom that they are not available for forming bonds, so that under no normal circumstances will hydrogen fluoride accept an additional proton to form H_2F^+; on the other hand, the proton in hydrogen fluoride itself can escape readily. Oxygen is less electro-negative than fluorine and water is a very much weaker acid (i.e. a proton escapes from the water molecule much less readily) than hydrogen fluoride; on the other hand, the two non-bonded pairs of electrons are available for forming bonds and water is a much stronger base than hydrogen fluoride so that H_3O^+ is readily formed. Ammonia is only a very weak acid and under no normal circumstances does it donate one of its hydrogen atoms in the form of a proton; on the other hand, its lone pair of electrons is readily available for forming a bond and ammonia is a very strong base. Finally, methane has no non-bonded electrons in its outer shell and the carbon–hydrogen bond is non-polar in character, as already discussed. In the alkyl halides the carbon–halogen bond was slightly polarized, but very much less so than in the hydrogen halides. We should therefore expect the extent of polarization in the alkyl–oxygen bond in alcohols and the alkyl–nitrogen bond in amines to be very small, i.e. there is *no* tendency for methanol to ionize as follows:

$$CH_3OH \not\longleftrightarrow CH_3{}^+ + OH^-$$

The main features of the chemistry of alcohols and amines is therefore not so much concerned with the carbon–oxygen bond in alcohols or the carbon–nitrogen bond in amines but rather with the reactions of the OH and NH$_2$ groups. In the first chapter and in the diagram on page 3, we have formally regarded methanol as water in which one hydrogen atom had been substituted by a CH$_3$ group and methylamine as ammonia in which one hydrogen atom had been substituted by a methyl group. We were solely concerned with indicating in a purely formal way how organic compounds can be regarded as being built up from familiar inorganic hydrides. The chemistry of methyl chloride has very little in common with the chemistry of hydrogen chloride and in this instance the relationship between methyl chloride and hydrogen chloride is purely formal. On the other hand, the chemistry of methanol and methylamine have a great deal in common with the chemistry of water and ammonia.

One of the most striking properties of water is its high boiling point (b.p. 100°C). Comparing this with the boiling point of ammonia (-30°C), methane (-162°C) or hydrogen sulphide (b.p. -61°C) shows how anomalous it is. One of the reasons for this high boiling point, and indeed one of the reasons for many of the other peculiarities of water, is the presence of hydrogen bonding. We can picture hydrogen bonding very qualitatively in the following way. One molecule of water readily transfers the proton to another molecule of water, so forming two ions OH$^-$ and H$_3$O$^+$. Although this 'autoprotolysis' occurs only to a very small extent it does indicate that the oxygen–hydrogen bond in water must be strongly polarized in the direction O$^{\delta-}$—H$^{\delta+}$. We can thus visualize a weak bond between the oxygen atom of one molecule and the hydrogen atom of another represented by .....
We make no attempt at the moment to describe the exact nature of this bonding beyond indicating that it may be of a polar nature:

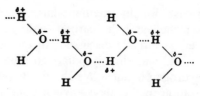

Clearly in an alcohol the opportunities for hydrogen bonding are reduced by half, but nonetheless it is still quite important and affects the physical properties of alcohol quite considerably. For example, the boiling points of simple alcohols are as follows: methanol, 65°C; ethanol, 78.5°C; propanol, 97°C. On the other hand, the boiling point of dimethyl ether, CH_3OCH_3, is -24°C and that of diethyl ether, $C_2H_5OC_2H_5$, is $+35$°C. Alcohols are thus all liquids, their boiling points increasing with the length of the carbon chain. The lower members of the series, methanol, ethanol and propanol, are completely miscible with water. The butanols are very soluble in water, but the higher members become increasingly less soluble as the number of carbon atoms increases. This gradation of properties is common to all homologous series (cf. Chapter 1).

The great feature of the chemistry of water is its amphoteric nature:

$HA + H_2O$	$\rightleftharpoons$	H_3O^+	$+ A^-$	Water as a base
		Hydroxonium ion		
$HA + C_2H_5OH$	$\rightleftharpoons$	$C_2H_5OH_2^+$	$+ A^-$	Alcohol as a base
		Ethyloxonium ion		
$B + H_2O$	$\rightleftharpoons$	$HB^+ +$	OH^-	Water as an acid
			Hydroxide ion	
$B + C_2H_5OH$	$\rightleftharpoons$	$HB^+ +$	$C_2H_5O^-$	Alcohol as an acid
			Ethoxide ion	

The acid dissociation constant of water is 3×10^{-16} Ethanol is a somewhat weaker acid, with $K_a = 10^{-18}$. The acidity of different alcohols differs only slightly though, in general, primary alcohols (RCH_2OH) are stronger acids than secondary alcohols (R_2CHOH), while tertiary alcohols (R_3COH) are the weakest. Since ethanol is a weaker acid than water, the sodium salt of ethanol cannot be made by treating it with sodium hydroxide; the easiest method for preparing sodium ethoxide is the direct reaction between sodium and ethanol. Sodium reacts with water so violently that the hydrogen evolved in the reaction may even ignite and explosions occur:

$$2\ Na + 2\ H_2O \longrightarrow 2\ Na^+ + 2\ OH^- + H_2(g)$$
(NaOH = sodium hydroxide)

With alcohol the reaction is very much less violent and alcohol can safely be treated directly with sodium:

$$2\ Na + 2\ C_2H_5OH \longrightarrow 2\ Na^+ + 2\ C_2H_5O^- + H_2\ (g)$$
(NaOC$_2$H$_5$ = sodium ethoxide)

The sodium salts of alcohols, called sodium *alkoxides*, are crystalline but they are very hygroscopic and hydrolyse in water to yield sodium hydroxide and the alcohol. The reaction of alkoxide anions with alkyl halides to yield ethers and a chloride anion was described in the last chapter:

$$RO^- \ \ \overset{\frown}{R'} - Cl \xrightarrow[\text{solvent}]{ROH} ROR' + Cl^-$$

Alkoxide   Alkyl   Ether
ion   halide

These reactions are carried out by using the alcohol from which the alkoxide ion is derived as solvent.

We noticed that the reaction of sodium with alcohol was exactly the same as the reaction of sodium with water except that it was much less violent. This is quite general; when an alcohol undergoes the same type of reaction as water, the alcohol reaction is usually much slower. Alcohol can replace water in many situations; e.g. salts which crystallize with water of crystallization often, but not always, also crystallize with alcohol of crystallization. The hydroxyl group of an oxygen acid can be replaced by the alkoxyl group of an alcohol to yield an ester. These are reactions which can actually be carried out, but for the moment we shall consider them as purely formal steps in the same way that we have already considered the building-up of the carbon chain. Thus, with sulphuric acid we have:

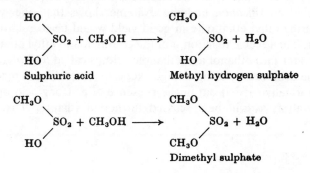

In a similar way a molecule of water can be eliminated from nitric acid and alcohol, giving ethyl nitrate and water:

$$\text{HO—NO}_2 + \text{C}_2\text{H}_5\text{OH} \longrightarrow \text{C}_2\text{H}_5\text{O—NO}_2 + \text{H}_2\text{O}$$
Nitric acid   Ethyl nitrate

Nitrate esters are explosive: glycerol trinitrate is manufactured under the name 'nitroglycerine':

$$\begin{array}{ccc}
\text{CH}_2\text{OH} & & \text{CH}_2\text{ONO}_2 \\
| & & | \\
\text{CHOH} + 3\,\text{HONO}_2 \longrightarrow & & \text{CHONO}_2 + 3\,\text{H}_2\text{O} \\
| & & | \\
\text{CH}_2\text{OH} & & \text{CH}_2\text{ONO}_2 \\
\text{Glycerol} & & \text{Glyceryl trinitrate}
\end{array}$$

a mixture of glycerol trinitrate and kieselguhr (finely divided SiO_2) constitutes dynamite.

A particularly important class of esters is that formed between alcohols and organic acids by elimination of water, e.g. ethyl ethanoate formed by elimination of water from ethanol and ethanoic acid:

$$\text{C}_2\text{H}_5\text{OH} + \text{HO—COCH}_3 \rightleftharpoons \text{C}_2\text{H}_5\text{O—COCH}_3 + \text{H}_2\text{O}$$
Ethanol   Ethanoic acid   Ethyl ethanoate

Notice that this is a reversible reaction; in other words, ethyl ethanoate reacts with water to regenerate ethanol and ethanoic acid. This particular reaction is a classical example of a reversible reaction.

$$K = \frac{[\text{CH}_3\text{COOC}_2\text{H}_5][\text{H}_2\text{O}]}{[\text{CH}_3\text{COOH}][\text{C}_2\text{H}_5\text{OH}]}$$

The equilibrium constant, K, at room temperature is here approximately 4. The reaction of an alcohol and an acid to form an ester is called *esterification* and the example shows that one way of preparing ethyl ethanoate in good yield would be to remove the water. The reverse reaction, i.e. the conversion of ethyl ethanoate and water into ethanol and ethanoic acid, is called *hydrolysis* and can be achieved by using a large excess of water. These reactions are normally carried out in the presence of a strong acid or base as catalyst, as will be discussed further in Chapter 9 (organic acids).

We began this chapter by emphasizing that the carbon–oxygen bond in alcohols was not very polar and that the principal reactions of alcohols were concerned with the OH group. We also mentioned how in the presence of a strong acid an alcohol can accept a proton, i.e. can behave as a base:

$$CH_3OH + HCl \rightleftharpoons CH_3\overset{+}{O}H_2 \quad + Cl^-$$
$$\textbf{Methyloxonium ion}$$

Now although the carbon–oxygen bond in methanol is only very slightly polar, the carbon–oxygen bond in the methyloxonium ion is clearly much more polar and it is possible for an anion to attack the carbon atom:

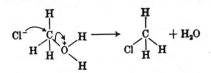

It would appear from the above two reactions that it would be possible to convert methanol into methyl chloride by treatment with hydrochloric acid. In fact this reaction does not occur in aqueous hydrochloric acid and is quite slow even when hydrogen chloride gas is dissolved in the methanol. However, by using phosphorus chlorides, PCl_3 or PCl_5, or thionyl chloride, $SOCl_2$, the acid chloride of sulphurous acid; the water formed in the reaction is completely removed at the same time as the hydrogen chloride is generated. This is represented formally in the equations

$$3 \text{ ROH} + PCl_3 \rightarrow 3 \text{ RCl} + H_3PO_3$$

$$ROH + SOCl_2 \rightarrow RCl + HCl + SO_2$$

The last reagent, $SOCl_2$, is particularly convenient as the inorganic products are gaseous and easily separated from the alkyl chloride.

Another way of obtaining the same result is to treat the alcohol with the sodium halide and sulphuric acid, the function of the acid being both to generate hydrogen chloride and so protonate the alcohol and to remove the water as it is formed; e.g. ethyl bromide can be prepared from ethanol, sodium bromide and sulphuric acid:

$$C_2H_5OH + Na^+ Br^- + 2 H_2SO_4 \longrightarrow$$
$$C_2H_5Br + Na^+ + H_3O^+ + 2 HSO_4^-$$

In discussing the reactions of alcohols, we have emphasized the importance of the hydroxyl group rather than the carbon–oxygen bond, excepting only the reactions discussed directly above in which displacement can occur with oxonium ion. Ethers have two carbon–oxygen bonds and no oxygen–hydrogen bond, and, in general, the reactions of ethers are similar to those of the alkanes (saturated hydrocarbons) in that they do not react readily with ionic species. However, just as it is possible to add an additional proton to the oxygen in an alcohol, so it is possible to add a proton to the oxygen in an ether molecule to given an oxonium ion:

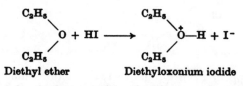

Diethyl ether   Diethyloxonium iodide

The oxonium ion is then capable of taking part in a displacement reaction, just as in the case of the oxonium ion from an alochol:

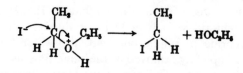

Diethyloxonium iodide   Ethyl iodide   Ethanol

Here the diethyloxonium ion is written in a form that illustrates how substitution by the donor, I^-, is identical in type with the reactions discussed in the previous chapter.

Ammonia is a gas (b.p. $-33°C$). The lowest member of the alkylamine series is methylamine, CH_3NH_2, which is a gas at normal temperature (b.p. $-7°C$). The other lower alkylamines are low-boiling liquids: dimethylamine, $(CH_3)_2NH$, b.p. 17°C; trimethylamine, $(CH_3)_3N$, b.p. 4°C; and ethylamine, $C_2H_5NH_2$, b.p. 17°C.

Ammonia is a base; it is completely miscible with water with which it reacts to form an ammonium cation and a hydroxide anion:

$$NH_3 + H_2O \rightleftharpoons NH_4^+ + OH^-$$

The alkylamines behave similarly although they are slightly stronger bases than ammonia; e.g. ammonia, $K_b = 2 \times 10^{-5}M$; methylamine, $K_b = 4 \times 10^{-4}M$; dimethylamine, $K_b = 5 \times 10^{-4}$M, where

$$K_b = \frac{[RNH_3{}^+][OH^-]}{[RNH_2]}$$

The alkylamines form stable crystalline salts with the mineral acids corresponding to the salts that ammonia forms, such as ammonium chloride or ammonium sulphate.

Amine salts		
$CH_3NH_3{}^+$ Cl^-	m.p. 226°C
$C_2H_5NH_3{}^+$ Br^-	m.p. 160°C
$(C_2H_5)_3NH^+$ Cl^-	m.p. 254°C

All these basic properties of the amines are due to the fact that the non-bonded pair of electrons on the nitrogen atom is readily available to form a bond. We have already discussed the importance of this in a different kind of reaction in connection with the reactions of the carbon–halogen bond. In the last chapter we discussed how the lone pair on the nitrogen atom of an amine would attack the carbon in an alkyl halide and eject the halide ion:

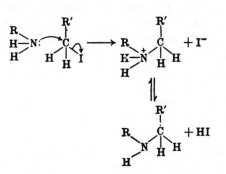

The nitrogen–hydrogen bond in ammonia or an alkylamine, unlike that in the ammonium ion or quaternary ammonium salt, does not separate readily as a proton, i.e. ammonia and alkylamines are very weak acids. Sodamide, NaNH$_2$, can be made by dissolving sodium in liquid ammonia, and similarly the sodium salts of amines can be prepared by the direct interaction of metallic sodium with the amine:

$$2\,RNH_2 + 2\,Na \longrightarrow 2\,R\bar{N}HNa^+ + H_2$$

However since alkylamines are stronger bases than ammonia, they must be even weaker acids, and this reaction only occurs very slowly.

Just as the OH group in oxygen acids can be replaced by the RO group of alcohols so it can also be replaced by the RNH or the R_2N group to yield amides; e.g. we can formally eliminate a molecule of water from methylamine and ethanoic acid to yield *N*-methylethanoamide:

$$CH_3NH_2 + HOCOCH_3 \longrightarrow CH_3NHCOCH_3 + H_2O$$

N-Methylethanoamide

At present we notice only the actual existence of these compounds; they will be discussed again in Chapter 9.

The Concept of a Functional Group

The last two chapters have surreptitiously introduced a basic concept in organic chemistry. This is the concept of a *functional group*. Particularly in the present chapter we have discussed the reactions of the OH group and the NH or the NH_2 groups, and have frequently disregarded the carbon chain. In the first chapter we discussed how long carbon chains can be built up; in the third chapter we described how the hydrocarbon chains were inert to ionic reactions. Thus an alkylamine reacts with water to form an aklylammonium ion and a hydroxide ion, regardless of the length or complexity of the alkyl chain or chains attached to the nitrogen atom. Similarly, an alcohol reacts with sodium to form a sodium alkoxide and hydrogen, regardless of the length or complexity of the carbon chain attached to the oxygen. For these reactions we treat the OH group and the NH or NH_2 groups as the 'functional groups' of the molecule. This concept is somewhat old-fashioned and can be confusing; nonetheless the idea that for ionic reactions, at least, only the functional group of an organic molecule is important can be useful. Even with the reactions of organic halides which concern reaction at a carbon atom, it can still be useful to regard the halogen as a functional group because the remainder of the carbon chain is usually unaffected by ionic reactions. For example, in the reaction between 1-chlorobutane and sodium ethoxide to yield butyl ethyl ether the carbon chain in the butyl group is completely unchanged by the reaction:

$$C_2H_5O^- \quad \overset{\overset{\textstyle C_3H_7}{|}}{CH_2} - Cl \longrightarrow C_2H_5OC_4H_9 + Cl^-$$

Nomenclature

The concept of a functional group is a cardinal principle in systematic nomenclature. Alcohols are given the name of the hydrocarbon from which they are derived, followed by the suffix '-ol'; e.g.

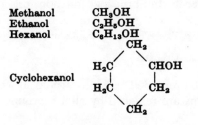

Methanol	CH_3OH
Ethanol	C_2H_5OH
Hexanol	$C_6H_{13}OH$
Cyclohexanol	

The word hexanol is clearly ambiguous and the position of the hydroxyl group is indicated by an arabic number; e.g.

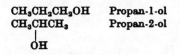

$CH_3CH_2CH_2OH$	**Propan-1-ol**
CH_3CHCH_3 $\quad\mid$ $\quad OH$	**Propan-2-ol**

For alcohols with less than five carbon atoms, semi-systematic names are frequently used. In the following scheme the systematic name is given with the semi-systematic name in square brackets:

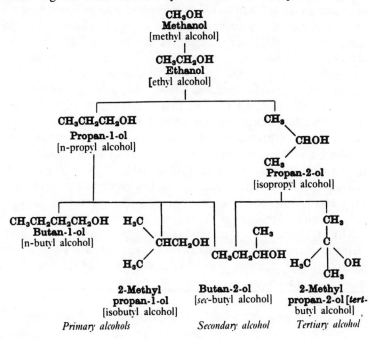

CH_3OH
Methanol
[methyl alcohol]

CH_3CH_2OH
Ethanol
[ethyl alcohol]

$CH_3CH_2CH_2OH$
Propan-1-ol
[n-propyl alcohol]

Propan-2-ol
[isopropyl alcohol]

$CH_3CH_2CH_2CH_2OH$
Butan-1-ol
[n-butyl alcohol]

2-Methyl propan-1-ol
[isobutyl alcohol]

Butan-2-ol
[sec-butyl alcohol]

2-Methyl propan-2-ol [tert-butyl alcohol]

Primary alcohols   *Secondary alcohol*   *Tertiary alcohol*

Alcohols are often classified as primary, secondary or tertiary according to atoms attached to the carbon atom adjacent to the hydroxyl group (RCH_2OH, primary; R_2CHOH, secondary; R_3COH, tertiary; where R is any alkyl radical).

A halogen-substituted alcohol is numbered so that the lowest number is given to the principal function, e.g.

$$CH_3CHCH_2CH—CHCH_3 \qquad \text{5-Chloro-3-methylhexan-2-ol}$$
$$\quad\; | \qquad\quad | \quad\; |$$
$$\quad\; Cl \qquad\;\; CH_3\; OH$$

Ethers can be regarded as hydrocarbons in which one or more hydrogen atoms are replaced by alkoxy groups. Thus

$$CH_3OCH_3 = \text{methoxymethane}$$

However, the older nomenclature by which the above compound is called dimethyl ether is still in general use, e.g.

$$CH_3CH_2OCH_3 = \text{methoxyethane or ethyl methyl ether}$$

The nomenclature of amines is somewhat exceptional. The correct systematic name for $CH_3CH_2NH_2$ would be ethanamine or 1-aminoethane, but by long-established custom radical names are attached to the ending amine, giving ethylamine for this compound. $(C_2H_5)_2NH$ is diethylamine and $(C_2H_5)_3N$ is triethylamine. Notice how nomenclature depends on the chemical properties. The principal reactions of alcohols involve breaking the oxygen–hydrogen bond and CH_3OH has quite different chemical properties to CH_3OCH_3 so that they receive quite different names (methanol for the former and methoxymethane for the latter). On the other hand, the reactions of amines depend principally on the availability of a non-bonded pair of electrons on the nitrogen so that the same kind of name is given to CH_3NH_2, $(CH_3)_2NH$ and $(CH_3)_3N$ (methylamine, dimethylamine and trimethylamine). For compounds containing quinquevalent nitrogen the ending '-mine' is changed to 'mmonium', that is $(CH_3CH_2)_4N^+I^-$ = tetraethylammonium iodide. The name changes as the chemical properties change. A tertiary amine with different alkyl chains is named as a derivative of the longest or most complex chain:

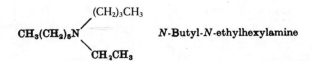

$$\text{CH}_3(\text{CH}_2)_5\text{N} \begin{array}{c} (\text{CH}_2)_3\text{CH}_3 \\ \\ \text{CH}_2\text{CH}_3 \end{array}$$ *N*-Butyl-*N*-ethylhexylamine

The italic capital *N* indicates that the substituent radicals are attached to the nitrogen. A compound containing both a hydroxy group and an amino group is regarded as an amino alcohol rather than a hydroxy amine, e.g.

$$\underset{\underset{\text{NH}_2}{|}}{\text{CH}_3\text{CHCH}_2\text{CH}_2}\underset{\underset{\text{OH}}{|}}{\text{CHCH}_3}$$ 5-Aminohexan-2-ol

Problems

1. What reaction, if any, would you expect between (1) ethanol and (2) ethylamine with the following reagents:

(a) HCl   (b) Na   (c) H$_2$O   (d) NaNH$_2$ (e) C$_2$H$_5$Br

2. What reaction, if any, would you expect between diethyl ether, C$_2$H$_5$OC$_2$H$_5$, and the following reagents:

(a) HBr   (b) Cl$_2$ (c) Na

3. Identify the compound that exhibits the following spectral characteristics:

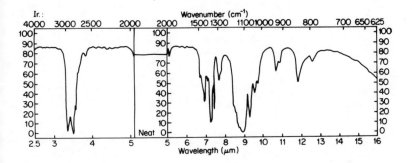

Problem 3 cont'd o/l

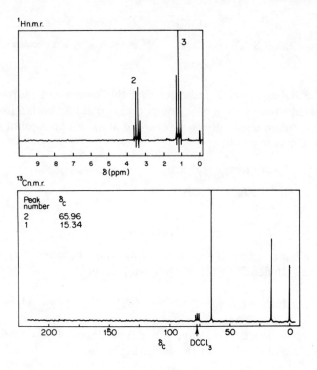

CHAPTER 6 ————————————

Elimination Reactions and the Formation of the Carbon–Carbon
Double Bond

We have represented the attack of an electron donor such as the
ethoxide ion, $C_2H_5O^-$, on an alkyl halide as follows:

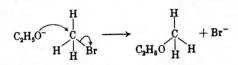

Ethoxide   Methyl bromide   Methyl ethyl ether
anion

When the alkyl halide is a methyl halide this is the only reaction
that can occur, but with the ethyl halides and the majority of
longer-chain alkyl halides another reaction can occur simultaneously; e.g.

$$C_2H_5O^- + CH_3CH_2Br \nearrow C_2H_5OC_2H_5 + Br^- \qquad \text{displacement}$$
$$\searrow CH_2{=}CH_2 + C_2H_5OH + Br^-$$
$$\textbf{Elimination}$$

In the second of these reactions the elements of hydrogen bromide
have been eliminated from ethyl bromide and this is therefore
called an *elimination reaction* in contrast to the displacement
described previously. This reaction can be represented as follows:

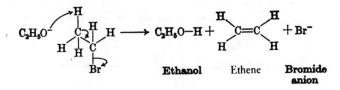

Ethanol   Ethene   **Bromide**
anion

71

Let us examine this reaction sequence very carefully. A pair of non-bonded electrons on the ethoxide anion is donated to one of the hydrogen atoms of the ethyl bromide, forming a new oxygen–hydrogen bond and ultimately a molecule of ethanol. The pair of electrons that originally formed the bond between the hydrogen atom and the carbon atom in the ethyl bromide are now transferred into the bond between the two carbon atoms. This means that the first carbon atom still retains eight electrons in its outer shell, and in order that the second carbon atom likewise only retains eight electrons in its outer shell the two electrons in the carbon–bromine bond must be transferred to the bromine atom to yield the bromide anion. We are thus left with a carbon–carbon bond in which four electrons are shared instead of two.

Before discussing the elimination any further, we must consider briefly the new type of molecule that has been formed in this reaction. All the chemical bonds between two atoms in the previous chapters have been formed by the sharing of two electrons. This type of bond can be called a *single bond*. In ethene the bond between the two carbon atoms is formed by the sharing of four electrons and this type of bond may reasonably be called a *double bond*. We shall see that the double bond undergoes many interesting reactions, but for the moment we will only consider its geometry. An important feature of the single bond, emphasized previously, is that there is free rotation between atoms along the axis of the bond and thus there are no positional isomers of, for example, 1,2-dichloroethane. The bonds in ethane subtend 109° 28′ to each other. In ethene all the atoms lie in one plane, the bonds subtending exactly 120° to each other. While it is possible to rotate atoms around a single bond this is not possible in a double bond. Thus there are two 1,2-dichloroethenes: *cis* and *trans*. The *cis*-isomer has the larger substituents on the same side of the molecule and the *trans*-isomer has the larger substituents on the opposite side of the molecule:

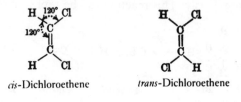

cis-Dichloroethene   *trans*-Dichloroethene

The *cis-* and *trans-*isomers are distinct compounds. Their physical properties are often very different (e.g. boiling points, melting points, etc.) and although their chemical reactions are necessarily similar there are some examples where the chemical reactivity can differ markedly.

Ethene is the simplest example of an *unsaturated compound* (i.e. one having a multiple bond). The reason for this name will be apparent in the next chapter when we come to consider the reactions of these compounds. The generic name for the unsaturated hydrocarbons is *alkene* (cf. alkane for saturated hydrocarbons). We can build up from ethene a series of alkenes by replacing hydrogen atoms severally by methyl groups, exactly as we built up the alkanes:

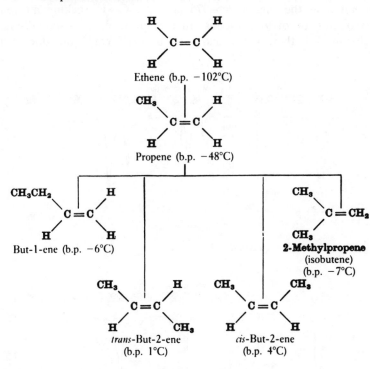

How many hydrocarbons can you draw with the formula C_5H_{10} (cf. Problem 1)?

The infrared spectra of alkenes with the double bond at the end of the molecule is substantially different from those with the double bond in the middle of a carbon chain. The first spectrum is of hex-1-ene (Figure 6.1). The peak a at ~ 3020 cm^{-1} (a) is the carbon–hydrogen stretching frequency of the vinyl group $=$C$-$H. The sharp peak b at ~ 1650 cm^{-1} is attributable to the carbon–carbon double bond stretching ($-$C$=$C$-$) and the strong broad peak c at 900 cm^{-1} is attributable to carbon–hydrogen out-of-plane bending.

In electronic spectra and particularly in n.m.r. spectra, the spectra of geometric isomers are similar. By their very nature it is clear that the infrared spectra of geometric isomers will be very different. The centre of mass of a *cis*-isomer will be different from that of the *trans*-isomer.

In the spectrum of *trans*-pent-2-ene (Figure 6.2) there is a strong bond a in the region 965–975 cm^{-1}, a C$-$H rocking motion characteristic of *trans*-olefins. In the spectrum of *cis*-pent-2-ene (Figure 6.3) there is a broad band b at 675 cm^{-1} also due to a

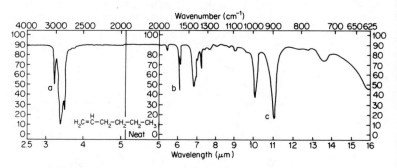

Figure 6.1 Infrared spectrum of hex-1-ene.

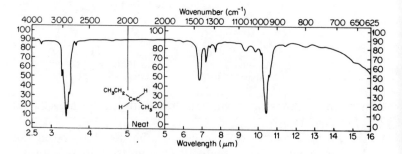

Figure 6.2 Infrared spectrum of *trans*-pent-2-ene.

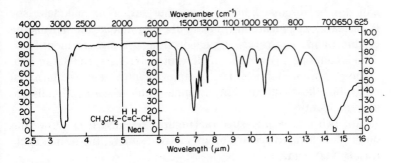

Figure 6.3 Infrared spectrum of *cis*-pent-2-ene.

C—H rocking motion (not always visible). The C=C absorption band (~1650 cm^{-1}) is quite strong in the *cis*-isomer but cannot be seen in the more symmetrical *trans*-isomer.

The first feature to notice about the ^{1}H n.m.r. spectrum of hex-1-ene (Figure 6.4) is the complex series of bands between δ 6.1 and 4.8 ppm (e). These are attributable to the vinyl hydrogen atoms (RCH=CH$_2$). The actual coupling is very complicated and

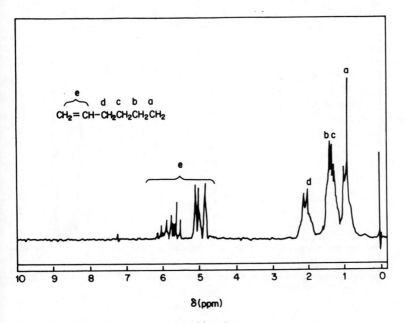

Figure 6.4 The ^{1}H n.m.r. spectrum of hex-1-ene.

will not be discussed here. The other important characteristic is the large chemical shift due to the deshielding of the alkene double bond. The remaining absorptions are easy to identify.

The coupling (d) of the vinylic protons (RCH=CHR') in hex-2-ene (Figure 6.5) are far less complex than those of the terminal olefin but the chemical shift is similar. The remaining bands are easy to identify.

The two ^{13}C n.m.r. spectra (Figures 6.6 and 6.7) show that unlike the infrared spectra of geometric isomers, the n.m.r. spectra are very similar.

We will return to the chemical reactions of the alkenes in the next chapter but at present we must consider eliminations in more detail. The first point to notice is that in Chapter 4 we said that the reaction between sodium ethoxide and ethyl bromide led to diethyl ether and now we have said that it may also lead to ethene. Great stress is usually laid on the percentage yield obtained in an organic reaction. Students sometimes disparagingly describe practical organic chemistry as cookery. When carrying out a preparation you are given a very detailed 'recipe' describing exactly

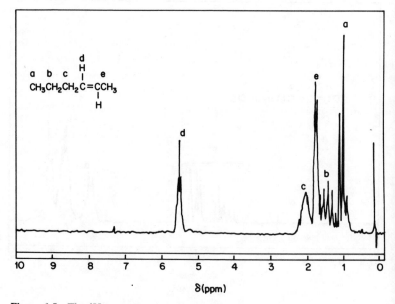

Figure 6.5 The ^1H n.m.r. spectrum of hex-2-ene.

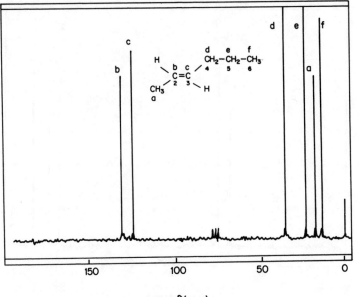

Figure 6.6 The ^{13}C n.m.r. spectrum of *trans*-hex-2-ene.

how many grams of this and how many grams of that you must mix together and how long you must heat them together, very similar to the directions in a cookery book for making a cake. The reasons why the directions have to be so explicit and why you have to record your yield is because in the reaction of most organic compounds with most reagents more than one reaction path is possible. In order to obtain the maximum yield of any desired compound, very particular conditions have to be adhered to. Let us look again at the elimination reaction as we have described it:

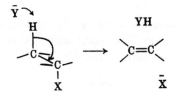

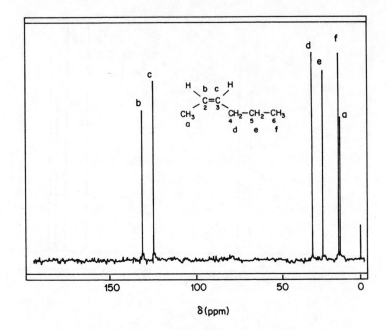

Figure 6.7 The ^{13}C n.m.r. spectrum of *cis*-hex-2-ene.

Notice that this process is analogous to the 'bimolecular displacement reaction'; we therefore describe this elimination as a 'bimolecular elimination reaction'. The two reactions usually occur side by side and in general the stronger the base the more important the elimination reaction becomes.

Strong bases   **$NH_2^- > OC_2H_5^- > OH^- > OCOCH_3^-$**   Weak bases

Elimination   Displacement

The structure of the alkyl halide also helps to determine which reaction predominates. The more branched the alkyl halide, the more likely the elimination reaction.

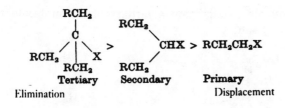

Tertiary   Secondary   **Primary**

Elimination   Displacement

The R in the above diagram can be H or any alkyl group.

With the above information we can now consider the best way in which to prepare methyl isopropyl ether, $CH_3OCH(CH_3)_2$. Suppose that 2-iodopropane was treated with sodium methoxide in methanol solution; from the above discussion we would expect the main products to be methanol, propene and sodium iodide:

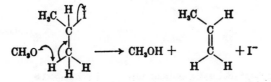

If, instead, methyl iodide was treated with sodium isopropoxide the principal product would be the desired ether:

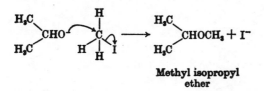

**Methyl isopropyl
ether**

In the eliminations described so far, the halide ions are called the *leaving group*. A good leaving group, i.e. a leaving group that facilitates elimination, is one which on elimination either yields a very stable ion (e.g. the halide ions already described) or else yields a stable molecule. Examination of the quaternary ammonium salts discussed in the previous two chapters shows that it is possible to eliminate a tertiary amine molecule from the quaternary salt:

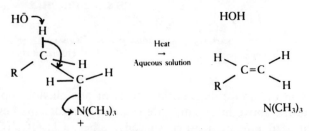

This is a very important reaction known as the *Hofmann elimination*. It is a quite general reaction and cycloheptyltrimethylammonium hydroxide can be decomposed to yield cycloheptene, trimethylamine and water:

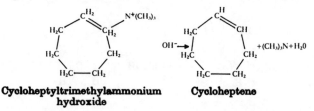

Cycloheptyltrimethylammonium   **Cycloheptene**
hydroxide

The Hofmann elimination gives us a clue to another type of very important elimination, namely, acid-catalysed dehydration of an alcohol. We know from the last chapter that alcohols behave as bases in the presence of a strong acid, to yield an oxonium ion:

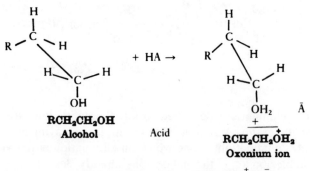

RCH₂CH₂OH   Acid   **RCH₂CH₂ȮH₂**
Alcohol   **Oxonium ion**

(or) $RCH_2CH_2OH + HA \rightarrow RCH_2CH_2\overset{+}{O}H_2\bar{A}$

The oxonium ion is structurally very closely related to the quaternary ammonium hydroxide discussed above, and elimination is therefore to be expected:

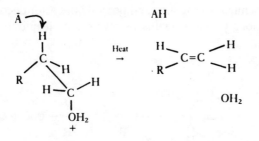

This dehydration of alcohols by strong acids is both common and important. Sometimes the reaction can be facilitated by making the acid derivative of the alcohol, i.e. the ester, and heating this. For example, ethanol can be dehydrated directly by sulphuric acid or alternatively ethyl hydrogen sulphate can be prepared at low temperatures, and this on heating yields ethene and regenerates sulphuric acid:

$$CH_3CH_2OH + H_2SO_4 \underset{}{\overset{0°C}{\rightleftarrows}} CH_3CH_2OSO_3H + H_2O$$

Ethyl hydrogen
sulphate
(ester)

$$CH_3CH_2OSO_3H \xrightarrow{Heat} CH_2{=}CH_2 + H_2SO_4$$

The esters of acetic acid can be pyrolysed at 400°C to yield olefins:

$$RCH_2CH_2OH + HOCOCH_3 \rightleftarrows RCH_2CH_2OCOCH_3 + H_2O$$

Ethanoic acid   Ethanoate (ester)

$$RCH_2CH_2OCOCH_3 \xrightarrow[400°C]{Pyrolysis} RCH{=}CH_2 + HOCOCH_3$$

At 400°C other kinds of decomposition are likely to set in and for this reason a more useful reaction is the *Tschugaev reaction* in which *xanthate* esters are pyrolysed:

$$RCH_2CH_2OH + CS_2 \xrightarrow{NaOH} RCH_2CH_2OCS_2{}^- + Na^+$$

Sodium xanthate
ester

$$RCH_2CH_2OCS_2^- \quad CH_3{-}I \longrightarrow RCH_2CH_2OCS_2CH_3 + I^-$$

Methyl xanthate
ester

$$RCH_2CH_2OCS_2CH_3 \xrightarrow[200°C]{Pyrolysis} RCH{=}CH_2 + COS + CH_3SH$$

The Tschugaev reaction, and possibly the ester decompositions, are reactions which probably involve the molecule 'biting its own tail', e.g.

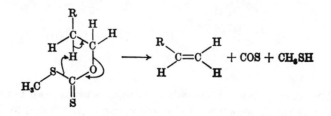

In the elimination reactions we have discussed so far reaction has been initiated by the attack of an anion (or by the electron donor 'tail' of the same molecule). These reactions correspond to the substitution reactions classified as displacement reactions and we call them 'bimolecular elimination reactions'. Just as there are substitution reactions initiated by the ionization of the reacting molecule so there are elimination reactions initiated in the same way; i.e. the rate-determining step of the process involves the departure of the leaving group and not the attack by an electron donor as in bimolecular elimination. Tertiary halides undergo unimolecular ionization rather than bimolecular displacements; analogously they undergo unimolecular elimination rather than bimolecular elimination:

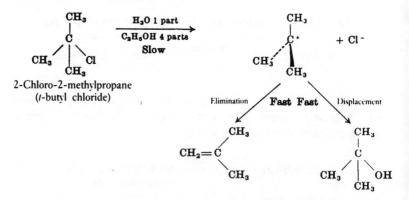

The rate of both these reactions depends only on the concentration of the 2-chloro-2-methylpropane.

Alkanes are also formed from vicinal dihaloalkanes by a variety of reactions. If, for example, 2,3-dibromobutane is treated with sodium iodide in acetone, instead of getting the expected 2,3-diiodobutane, but-2-ene is formed:

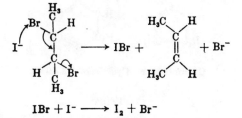

$$IBr + I^- \longrightarrow I_2 + Br^-$$

This reaction is usually represented as shown, i.e. as a bimolecular elimination reaction in which the iodide anion abstracts Br^+ from the dibromide. Two adjacent halogen atoms can also be removed by treatment with metals such as zinc:

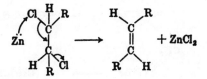

This reaction undoubtedly goes by the same process as a bimolecular elimination reaction.

Nomenclature

Unsaturated, unbranched acyclic hydrocarbons having one double bond are named by replacing the ending '-ane' of the corresponding saturated hydrocarbon by the ending '-ene'. If there are two or more double bonds, the ending will be '-adiene', '-atriene', etc. The chain is so numbered to give the lowest possible numbers to the double bonds; e.g.

$CH_3CH_2CH_2CH_2CH=CH_2$ Hex-1-ene
$CH_3CH_2CH_2CH=CHCH_3$ Hex-2-ene
$CH_3CH=CHCH_2CH=CH_2$ Hexa-1,4-diene
$CH_2=CHCH=CHCH=CH_2$ Hexa-1,3,5-triene

The generic name of these hydrocarbons is *alkene* (alkadiene, alkatriene, etc.).

Monocyclic compounds are named in the same way, e.g.

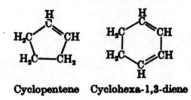

Cyclopentene Cyclohexa-1,3-diene

Problems

1. How many alkenes with molecular formula C_5H_{10} can you draw? (Include *cis*- and *trans*-isomers.)

2. Suggest reaction sequences for the following transformations:

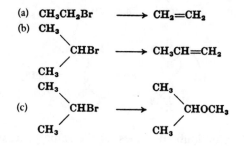

3. Suggest a structure for a compound possessing the following spectral data:

MS: M^+ 84

^{13}C decoupled δ 24 ppm; δ 35 ppm; δ 114 ppm; δ 144 ppm.

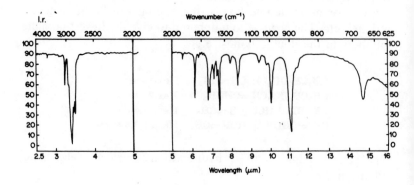

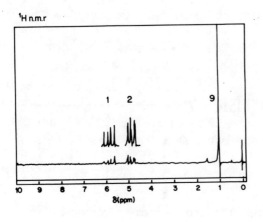

CHAPTER 7

Addition Reactions: to the Carbon–Carbon Double Bond

The energy required to break a carbon–carbon single bond is about 348 kJ mol^{-1}:

$$\overset{\displaystyle |}{\underset{\displaystyle |}{C}}\diagdown\overset{\displaystyle |}{\underset{\displaystyle |}{C}}\diagup \longrightarrow \overset{\displaystyle |}{\underset{\displaystyle |}{C}}\cdot \;+\; \overset{\displaystyle |}{\underset{\displaystyle |}{C}}\cdot \qquad \varDelta H \approx +348 \text{ kJ mol}^{-1}$$

The uncoupling of one of the pairs of electrons in a double bond, however, only requires about 264 kJ:

$$\underset{\diagup}{\overset{\diagdown}{C}}{=}\underset{\diagdown}{\overset{\diagup}{C}} \longrightarrow -\underset{\displaystyle \cdot}{C}-\underset{\displaystyle \cdot}{C}- \qquad \varDelta H \approx +264 \text{ kJ mol}^{-1}$$

It follows that 'addition reactions' are very exothermic processes:

$$CH_2{=}CH_2 + Br_2 \longrightarrow CH_2BrCH_2Br \quad \varDelta H \approx -109 \text{ kJ mol}^{-1}$$
$$CH_2{=}CH_2 + H_2 \longrightarrow CH_3CH_3 \qquad\;\; \varDelta H \approx -126 \text{ kJ mol}^{-1}$$

The above equations indicate that the addition of bromine or of hydrogen to ethene is an exothermic process, i.e. it is downhill energetically. This does not mean, however, that either of these reactions will occur in the fashion described in the above equations. A mixture of gaseous ethene and gaseous hydrogen in a flask would remain unchanged indefinitely, and even a mixture of gaseous bromine and gaseous ethene in a completely clean flask in the dark would remain unchanged for a considerable time. In Chapter 3 we saw that, although the chlorination of methane was an exothermic process, methane and chlorine do not react until light is shone on the gaseous mixture. All we learn from knowing that a reaction is very exothermic is that if we can find a way of initiating it, it is likely to continue readily.

Addition reactions to double bonds, i.e.

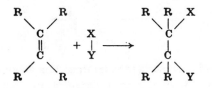

can go by four different mechanisms.

1. Addition of electron acceptors to alkenes (very common and important for hydrocarbon alkenes):

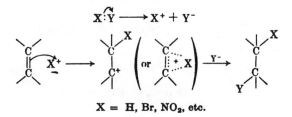

$$X = H, Br, NO_2, etc.$$

X^+ is an electron acceptor which needs a pair of electrons to complete its outer shell. We must compare an electron acceptor with an electron donor (Chapter 4). An electron acceptor need not necessarily carry a positive charge but it must be a species which requires electrons to complete its outer shell (e.g. boron trifluoride, BF_3). In a similar way not all electron donors carry a negative charge but they have a non-bonded pair of electrons (e.g. trimethylamine, $(CH_3)_3N:$). In the above reaction sequence one of the pairs of electrons in the carbon–carbon double bond is donated to the electron acceptor. If X^+ is a proton, a new carbon–hydrogen bond is formed, but for other electron acceptors we have a triangular structure with dotted lines from the two carbon atoms to X. This is to represent that the electron acceptor is not attached specifically to one carbon atom but is bonded to some extent to both. There is good evidence that with the addition of an electron acceptor to carbon–carbon double bonds a bridged intermediate complex is formed. The same is not true for other types of additions.

2. Addition of electron donors (uncommon for hydrocarbon alkenes):

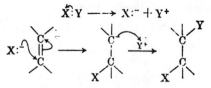

Electron donors do not add to hydrocarbon alkenes and we shall, therefore, not be discussing this kind of addition further in this chapter; however in subsequent chapters we shall show that for carbon–oxygen double bonds it is a very important reaction.

3. Free-radical addition:

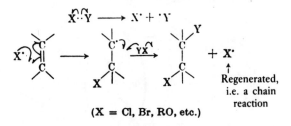

(X = Cl, Br, RO, etc.)

In the above scheme, we have introduced a new symbolism. In Chapter 4 we introduced the use of a curved arrow with a full head to represent the transfer of a pair of electrons from one nucleus to another (⌢). We now introduce a curved arrow with only half a head (⌢). This half-headed arrow is used to represent the *transfer of a single electron* from one nucleus to another.

4. Pericyclic reactions:

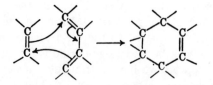

The term 'pericyclic reaction' refers to a reaction in which there is a redistribution of electrons in a cyclic system and the nuclei undergo valency changes more or less simultaneously.

The additions (shown in 1 and 2) of the electron acceptor and of the electron donor are *heterolytic* reactions and the addition shown in 3 (free-radical addition) is a *homolytic* addition (cf. Chapter 3).

Addition of Electron-pair Acceptors to Alkenes

Mineral Acids

Mineral acids add across carbon–carbon double bonds in the hydrocarbon alkenes:

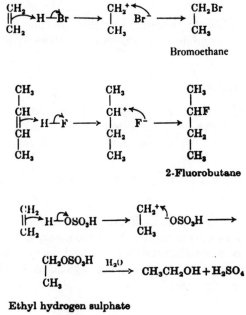

Bromoethane

2-Fluorobutane

Ethyl hydrogen sulphate
(ester)

The last of the above reactions, that of ethene with sulphuric acid, is a very important industrial process. Ethanol used to be prepared solely by fermentation but the demand now far exceeds the possible supply by this means. Industrial alcohol is therefore also prepared by the addition of sulphuric acid to ethene. The product of this addition, ethyl hydrogen sulphate, is an ester of sulphuric acid and ethanol. In Chapter 5 an ester was defined as a molecule of an acid from which the elements of water had been replaced by the elements of ethanol; this reaction is, in general, reversible, so that if ethyl hydrogen sulphate is treated with an excess of water, sulphuric acid and ethanol will be formed.

Addition of Halogens

Bromine and chlorine add in a heterolytic fashion if the reaction is carried out in the liquid phase in the dark.

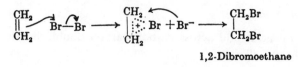

1,2-Dibromoethane

We shall see below that bromine and chlorine can also add by a homolytic process and fluorine adds only by a homolytic reaction.

Addition of 'Hypohalous Acids'

When bromine or chlorine is dissolved in water, a solution called bromine water or chlorine water is obtained and this solution contains some hypohalous acid:

$$Br_2 + H_2O \rightleftharpoons BrOH + HBr$$

The experimental fact is that a hydrocarbon alkene, treated with the solution, adds the elements of the hypohalous acid. However, there is very good experimental evidence to show that the reaction does not involve addition of the hypohalous acid itself but involves the following mechanism:

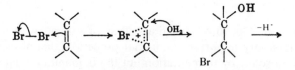

If the alkene is treated with a solution of bromine in ethanol, the products of the reaction are the 2-bromo ether, or if the reaction is carried out in an aqueous solution containing another anion then the product contains the other anion, as shown in the following equations:

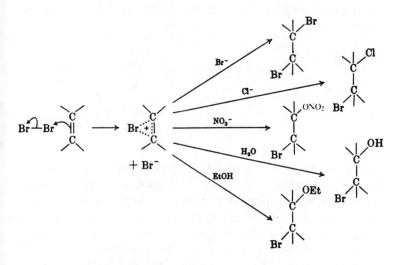

Addition of Peroxy Acids

When a carboxylic acid is treated with a solution of hydrogen peroxide the —OH group is replaced by an —OOH group. The product of this reaction is called a peroxy acid:

$$CH_3CO—OH + H_2O_2 \rightleftharpoons CH_3CO—O_2H + H_2O$$
(ethanoic acid)   perethanoic acid

$$HCO—OH + H_2O_2 \rightleftharpoons HCO—O_2H + H_2O$$
(methanoic acid)   methanoic acid

In practice the peroxy acid is not isolated. The solution of the carboxylic acid in aqueous hydrogen peroxide is used directly, e.g. for reaction with alkenes:

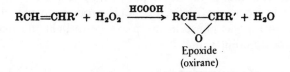

Epoxide
(oxirane)

The product of this reaction is a cyclic ether, called an *epoxide*; the systematic name for such ring ethers is *oxirane*. The mechanism of the addition involves ethene as the donor like the reactions discussed above:

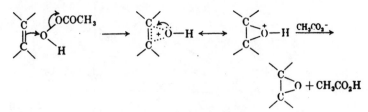

The initial stages of this reaction are probably identical with all the electron acceptor additions so far discussed. The reaction is initiated by the electrons in the oxygen–oxygen bond separating and leaving the OH group of the peroxy acid with a positive charge. This positive charge is neutralized by a pair of electrons from the double bond to give us the same type of addition complex as we had above. What is different about this addition complex is that the oxygen atom has two non-bonded pairs of electrons, one of which can be used to form a bond with the carbon atoms, so that the positive charge is no longer distributed over the two carbon atoms and the oxygen but is now located on the oxygen atom, thus giving two normal bonds between the oxygen atom and the two carbon atoms. We represent the transition from one form of the addend complex to the other by a double-headed arrow to indicate that this reaction only involves the movement of electrons and not the movement of nuclei. The use of a double-headed arrow to represent structures which differ only in the position of electrons will be discussed in much greater detail later on.

Homolytic Additions

We have seen that in the liquid phase in the dark, chlorine and bromine add by a heterolytic, i.e. ionic process, to the olefins. However, in the gas phase in the presence of light the reaction occurs by a different route:

$$Cl_2 + h\nu \xrightarrow{\ 1\ } 2\,Cl^{\cdot}$$

$$Cl^{\cdot} \quad CH_2{=}CH_2 \xrightarrow{\ 2\ } Cl{-}CH_2{-}CH_2^{\cdot}$$

$$Cl{-}CH_2{-}CH_2^{\cdot} \quad Cl{-}Cl \xrightarrow{\ 3\ } Cl{-}CH_2{-}CH_2{-}Cl + Cl^{\cdot}$$

$\Big\}$ Chain reaction

This is another example of a chain reaction (see Chapter 3). Reactions 2 and 3 are the chain propagating steps. The chain terminating steps are the same as those for the chlorination of methane as given in Chapter 3. Apart from the fact that this free-radical reaction proceeds most readily in the gas phase there is another important difference. The addition of bromine to propene in the liquid phase in the dark yields only a single product:

$$Br_2 + CH_3CH{=}CH_2 \xrightarrow[\text{in the dark}]{\text{Liquid phase}} CH_3CHBrCH_2Br$$

1,2-Dibromopropane
(only product)

This is because Br^+ will not attack a carbon–hydrogen bond (see Chapter 3). On the other hand, a bromine atom will abstract a hydrogen atom from a carbon–hydrogen bond:

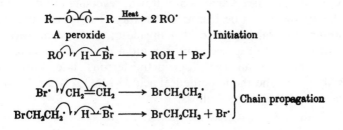

Thus bromine and propene react together in the gas phase in the presence of light to yield three products: 1,2-dibromopropane (as before), 3-bromoprop-1-ene and 1,2,3-tribromopropane. The tribromo compound comes from the addition of the bromine to the bromopropene formed by the abstraction reaction. Homolytic additions are not restricted to halogen atoms, and it is possible to add such apparently ionic molecules as hydrogen bromide homolytically, although reactions of this kind require an initiator:

$$R{-}O{-}O{-}R \xrightarrow{\text{Heat}} 2\,RO^{\cdot} \left.\begin{array}{c} \\ \\ \end{array}\right\}\text{Initiation}$$

A peroxide

$$RO^{\cdot} + H{-}Br \longrightarrow ROH + Br^{\cdot}$$

$$Br^{\cdot} + CH_2{=}CH_2 \longrightarrow BrCH_2CH_2^{\cdot} \left.\begin{array}{c} \\ \\ \end{array}\right\}\text{Chain propagation}$$

$$BrCH_2CH_2^{\cdot} + H{-}Br \longrightarrow BrCH_2CH_3 + Br^{\cdot}$$

In this reaction the initiator is a peroxide in which the oxygen–oxygen bond is broken homolytically by heat.

Oxidation Reactions (Pericyclic Reactions)

Carbon–carbon double bonds react with powerful oxidizing agents such as potassium permanganate, osmium tetroxide or ozone to yield addition products. The mechanism of these reactions is very complex and numerous side reactions occur. Nonetheless the basic reaction in each case is probably essentially the same.

Potassium Permanganate

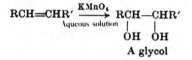

$$RCH{=}CHR' \xrightarrow[\text{Aqueous solution}]{KMnO_4} RCH{-}CHR'$$
$$\underset{\text{OH}}{|} \quad \underset{\text{OH}}{|}$$

A glycol

This reaction is probably initiated by a four-centre reaction between two of the oxygen atoms of the permanganate ion and the alkene:

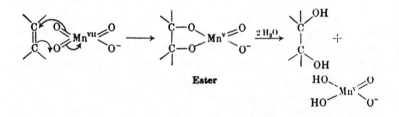

Ester

Notice that in this particular reaction it does not matter which way round the curved arrows are drawn, reversing the directions of the arrows would be just as satisfactory. The roman numerals represent the valency state of manganese; in the same way we can write Fe^{III} to represent an iron atom in the ferric state and Fe^{11} to represent an iron atom in the ferrous state. Mn^V is an unstable state and the hypothetical acid $H_2MnO_4{}^-$ will react with a permanganate ion as follows:

$$H_2Mn^VO_4{}^- + Mn^{VII}O_4{}^- \longrightarrow 2\,HMn^{VI}O_4{}^-$$
Manganate anion

Osmium Tetroxide

An alkene reacts with osmium tetroxide to produce an osmate ester which will hydrolyse to yield a glycol in a way analogous to the reaction of an alkene with permanganate. In this case the mechanism of the reaction is far more certainly known and the intermediate osmate ester can be isolated:

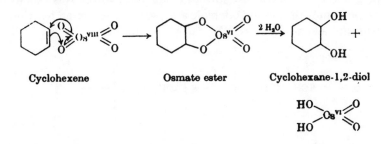

Cyclohexene   **Osmate ester**   **Cyclohexane-1,2-diol**

As osmium tetroxide is an expensive (and also poisonous) reagent it can be regenerated using a cheaper oxidant such as sodium chlorate, allowing only a catalytic amount of osmium tetroxide to be used to effect the oxidation:

$$ClO_3^- + 3(HO)_2Os^{VI}O_2 \rightarrow Cl^- + 3H_2O + Os^{VIII}O_4$$

Ozone probably adds to the alkene in exactly the same way as permanganate and osmium tetroxide, but the initial product is unstable and complete fission of the carbon–carbon bond occurs to yield an ozonide, a five-membered ring containing three oxygen atoms:

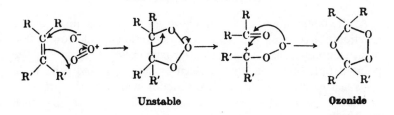

Unstable   **Ozonide**

The ozonides are very reactive substances and can be hydrolysed by water:

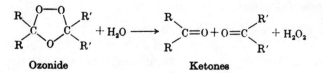

Ozonide   **Ketones**

Because the product of the reaction with water yields hydrogen peroxide which may cause further oxidation, hydrolysis is usually carried out with zinc and ethanoic acid, so that the overall reaction with the alkene RCH=CHR' can be written as

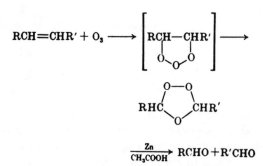

Triphenyl phosphine (and also dimethyl sulphide) can be used in place of zinc and ethanoic acid:

$$Ozonide + (C_6H_5)_3P \rightarrow 2R_2CO + 2(C_6H_5)_3PO$$
$$\text{Triphenyl}$$
$$\text{phosphine}$$

Hydrogenation Reactions

At the beginning of this chapter we set out an equation for the reaction of hydrogen with ethene to yield ethane and showed that this reaction was exothermic to approximately 126 kJ mol^{-1}. We pointed out, however, that molecular hydrogen and ethene do not react with one another. This reaction will go in the presence of certain metals which acts as catalysts. The exact way in which the catalyst acts is not completely understood but we do know that alkenes are absorbed quite strongly on the metal surface. Presumably the surface of the metal contains unsatisfied valencies which form bonds with the alkene. Molecular hydrogen is also absorbed on metal surfaces although the metal–hydrogen bond is considerably weaker. It seems probable that the actual hydrogenation occurs by reaction between absorbed alkene and absorbed hydrogen:

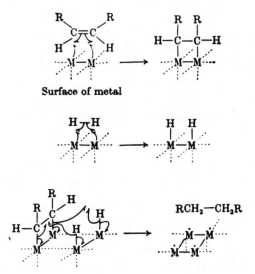

Notice that in this process the catalyst is regenerated and so the metal acts as a true catalyst, i.e. only a very small area of metal surface is required.

Another method of reduction of the double bonds involves hydroboration. This is an example of addition by an electron acceptor. The first step is the addition of diborane to an alkene:

$$6 \ CH_2{=}CH_2 + B_2H_6 \xrightarrow{0°C} 2 \ (CH_3CH_2)_3B$$

Diborane is spontaneously inflammable in air and in practice it is formed in solution by treating sodium borohydride with boron trifluoride:

$$3 \ Na^+BH_4^- + 4 \ BF_3 \ (in \ Et_2O) \xrightarrow{0°C} 2 \ B_2H_6 + 3 \ Na^+BF_4^-$$

Boron compounds can be electron acceptors in the same way as nitrogen compounds can be electron donors. In a trivalent boron compound the boron atom has only six electrons in its outer shell. Boron trifluoride and ammonia react to form a complex which has a very strong bond between the nitrogen and the boron. Two non-bonded electrons from the nitrogen atom complete the outer shell of the boron. The addition of diborane to an alkene starts by the donation of a pair of electrons from the carbon–carbon double bond to the boron:

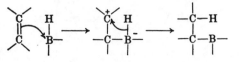

The alkylborane can then be decomposed by aqueous acid to yield the hydrocarbon or with hydrogen peroxide and alkali to yield an alcohol:

$$(CH_3CH_2)_3B + 3 H_2O \xrightarrow{H^+} 3 CH_3CH_3 + B(OH)_3$$
$$(CH_3CH_2)_3B + 3 H_2O_2 \xrightarrow{OH^-} 3 CH_3CH_2OH + B(OH)_3$$

Polymerization

At the beginning of this chapter we considered the types of addition that could occur at a double bond and the first three that we listed were:

1. Addition of electron acceptors to alkenes
2. Addition of electron donors to alkenes
3. Addition of free radicals to alkenes

The initial product in the addition of an alkene to an electron acceptor must itself be an electron acceptor, so it would be reasonable to expect it to add to another alkene molecule, the probability of such a reaction depending on the concentration of the alkene:

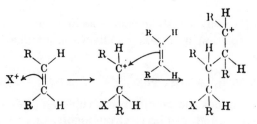

If we treat 2-methylpropene with sulphuric acid, for example, we obtain two dimers by this process:

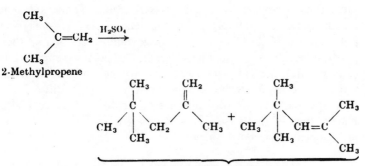

2-Methylpropene

Dimers of 2-methylpropene

If instead we use BF_3 as the electron acceptor there is no anion to halt the reaction and a long-chain polymer is obtained:

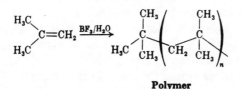

Polymer

This kind of polymerization initiated by an electron acceptor is called *cationic polymerization*.

The fact that addition of electron donors to hydrocarbon alkenes is rare has been discussed and thus *anionic polymerization* in which the polymerization process is initiated by an anion and the growing chain is a carbanion is likewise not important for hydrocarbon alkenes, though anionic polymerization can occur with other types of molecules.

The third kind of addition, free-radical addition, is very important for the polymerization of hydrocarbon alkenes:

$$X^{\cdot} \; CH_2 \!=\! CH_2 \longrightarrow XCH_2 - \dot{C}H_2 \xrightarrow{CH_2 = CH_2}$$

$$XCH_2 - CH_2 - CH_2 - \dot{C}H_2 \xrightarrow{n-2 \; CH_2 = CH_2} X - (CH_2 - CH_2)_n$$

It can be seen that this reaction and ionic polymerization are chain reactions and in order for the chains to be long to give a long polymer we must have a high concentration of ethene relative to the initiator. Poly(ethene) known as 'alkathene' or 'polythene' is probably the most common plastic today. The ordinary polythen used for packaging or for the very numerous toys or kitchen utensils is prepared by exactly this reaction. There are other alkene polymers, notably poly(propylene) ($CH_3CH=CH_2$) and poly(styrene) ($C_6H_5CH=CH_2$). A large number of other synthetic plastics and fibres are polymers. Some, like poly(styrene) (a hard clear plastic), are very similar to poly(ethene) in basic structure. Others, like nylon (to be discussed in Chapter 18), differ in that they are made up of two different molecules which occur alternately, that is A—B—A—B—, instead of the A—A—A—A— arrangement as in poly(ethene).

Unsaturation

In the previous chapter we stated that ethene is the simplest example of an unsaturated compound. Alkenes are given this name becuase they undergo addition reactions, and the whole of this chapter has been devoted to describing the principal types of addition reaction. The alkanes, the reactions of which were discussed in Chapter 3, are called saturated hydrocarbons and you will remember that the saturated bonds in a hydrocarbon do not undergo ionic reactions. Alkenes, on the other hand, undergo a wide variety of ionic and homolytic reactions. The next few chapters will be concerned with further examples of addition reactions—not to carbon–carbon double bonds but to carbon–oxygen double bonds. The term unsaturation is somewhat loosely used, but according to many authors it only refers to carbon–carbon bonds; carbon–oxygen bonds would not be regarded as unsaturated by these authors. We think this is an unfortunate distinction and it is one that we will endeavour to avoid using.

Problems

1. Starting from ethene, how would you prepare the following compounds:

(a) C_2H_5OH (b) CH_2BrCH_2OH (c) $CH_2(OH)CH_2OH$
(d) CH_3CH_2Br (e) $—(CH_2CH_2)_n—$ poly(ethene)

2. Distinguish between homolytic and heterolytic addition. Show how bromine can add either heterolytically or homolytically to but-2-ene. Under what conditions would you expect heterolytic addition and under what conditions would you expect homolytic addition? What additional products might you expect from the homolytic reaction?

CHAPTER 8

Addition Reactions: to the Carbon–Oxygen Double Bond

The previous two chapters have been concerned with bonds between carbon atoms in which four electrons are shared, giving a double bond. We can equally well form double bonds between carbon and nitrogen or carbon and oxygen atoms:

Imine group   Carbonyl group

The infrared spectra of pentan-2-one (methylpropyl ketone) (Figure 8.1) and 1-pentanal (valeraldehyde) (Figure 8.2) both have strong bands at 1710–1720 cm^{-1} due to the carbonyl stretching frequency. The lone hydrogen atom attached directly to the carbonyl in aldehydes gives two bands, 2740 and 2855 cm^{-1}, not apparent in the isomeric ketones.

The ^{1}H n.m.r. spectrum of pentan-2-one (Figure 8.3) shows a sharp singlet at b ($\delta \sim 2.2$ ppm) attributable to the isolated CH$_3$

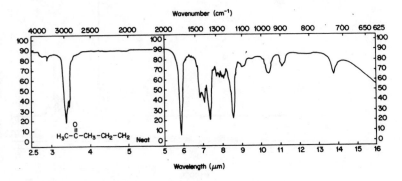

Figure 8.1 Infrared spectrum of pentan-2-one.

101

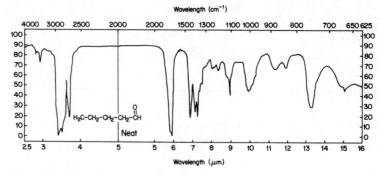

Figure 8.2 Infrared spectrum of pentanal.

group. The absorption due to the other CH_3 group is a triplet centred around δ 0.90 ppm at d, split by the adjacent CH_2 group. There is a second triplet at a (δ 2.45 ppm) attributable to the CH_2 group adjacent to the carbonyl. The remaining peak at c (*circa* δ 1.6 ppm) is a multiplet from coupling with both the terminal methyl group and the central CH_2 group.

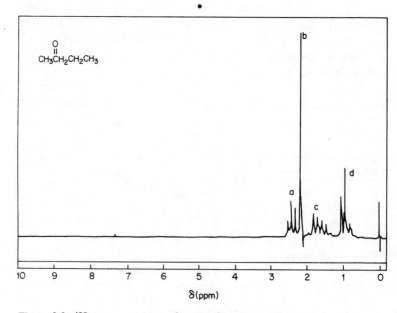

Figure 8.3 ^1H n.m.r. spectrum of pentan-2-one.

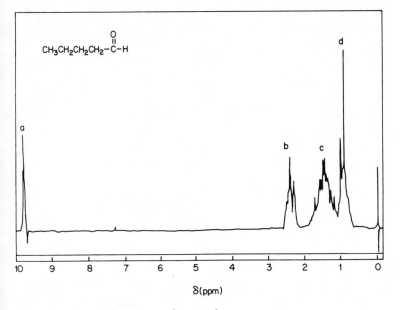

Figure 8.4 ^{1}H n.m.r. spectrum of pentanal.

The striking feature of the pentanal spectrum (Figure 8.4) is the sharp multiplet at δ 9.8 ppm (a) due to the hydrogen bound to the carbonyl. The remainder of the spectrum consists simply of a terminal CH_3 group d, a complex multiplet (δ 1.49 ppm) c due to four protons and a badly resolved triplet (δ 2.45 ppm) b due to two protons.

The ^{13}C n.m.r. spectrum of pentan-2-one (Figure 8.5) shows one carbon atom well separated from the rest of the spectrum a (δ_C 207.75 ppm) which can be attributed to the carbonyl carbon. The carbon atoms adjacent to the carbonyl can be associated with the peaks at b (δ_C 45.58 ppm) and c (δ_C 29.69 ppm) and the remaining peaks at d (δ_C 17.53 ppm) with the CH_2 group and at e (δ_C 13.81 ppm) with the terminal CH_3 group.

This chapter will not be concerned with the carbon–nitrogen double bond (as occurring in the imine group) but with the carbon–oxygen double bond (as in the carbonyl group), which is one of the most important groups in organic chemistry. The simplest molecule containing a carbonyl group is methanal and it is possible to build up a series of compounds containing the carbonyl group by successively replacing hydrogen atoms by

methyl groups in just the same way as we built up the alkanes
and the alkenes:

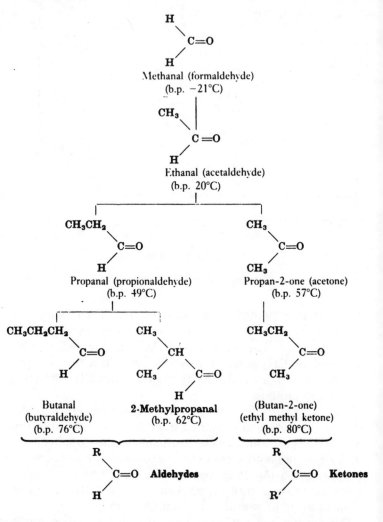

Notice that so long as a hydrogen atom is still attached directly
to the carbon atom of the carbonyl group, the compound is called
an aldehyde. The addition reactions of aldehydes and ketones
with which we are concerned at present are the same.

In Chapter 4 we considered the single bond between carbon
and halogen in an alkyl halide and came to the conclusion that

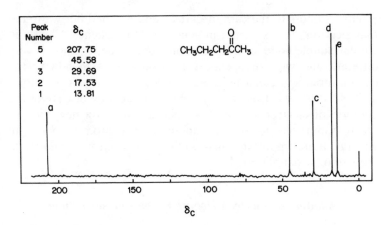

Peak Number	δ_C
5	207.75
4	45.58
3	29.69
2	17.53
1	13.81

$$CH_3CH_2CH_2\overset{\overset{\displaystyle O}{\|}}{C}CH_3$$

Figure 8.5 ^{13}C n.m.r. spectrum of pentan-2-one.

the pair of electrons would not be equally shared between the carbon and the halogen atom and the bond is therefore slightly polar in the direct ion C+→Cl:

We thus distinguished between the completely non-polar carbon–carbon single bond and the slightly polar carbon–halogen single bond. In exactly the same way the carbon–carbon bond in ethene is completely non-polar, the four electrons being equally shared between the two carbon atoms. In a carbon–oxygen double bond we may expect the more electronegative oxygen atom to have slightly the greater share of the electrons and the double bond will therefore be polar:

If we now compare the carbon–oxygen double bond with the carbon–carbon double bond we should expect the polarity of the former to make ionic addition occur more readily. In the previous chapter we considered the addition of both electron acceptors and electron donors to the carbon–carbon double bond but we found that electron donors do not in general add to alkenes. Clearly in a polar double bond as in the carbonyl bond, the situation is very different and we would expect electron donors to add to the carbon atom and electron acceptors to add to the oxygen atom. This is exactly what does occur. We will begin by discussing the addition of electron donors first, because this is a new kind of reaction and, second, because addition of electron donors is, in general, more important than addition of electron acceptors to the carbonyl double bond.

Addition of Electron Donors to the Carbonyl Bond

The commonest electron donor is, of course, the hydroxyl anion, which will add to the carbonyl bond as shown:

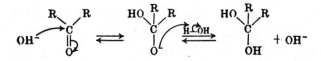

Notice that this addition amounts to the addition of water to the carbonyl bond and that the products of the reaction contain two hydroxyl groups on the same carbon atom. This is not a stable molecule in most cases, and notice that the reaction has been drawn as being reversible. In some special cases the hydrate, as the reaction product is called, can be quite stable. For instance, trichloroethanal, called chloral, forms an exceedingly stable hydrate, called chloral hydrate:

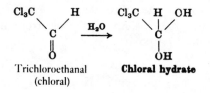

Trichloroethanal   **Chloral hydrate**
(chloral)

In general, however, the aldehyde and ketone hydrates are unstable substances and exist only in solution.

In the substitution reactions of alkyl halides, an important electron donor was the cyanide anion. The cyanide anion will add to the carbonyl bond:

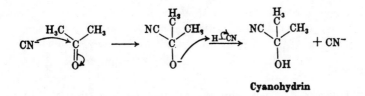

Cyanohydrin

This reaction is carried out in an aqueous solution containing an excess of sodium cyanide to which one mole of acid has been added. The reaction is effectively the addition of hydrogen cyanide across the carbon–oxygen double bond. Another example of the anion of a weak acid adding to the carbonyl bond is the bisulphite anion:

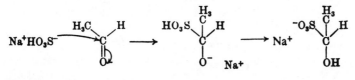

Sodium bisulphite

This reaction occurs much more readily with aldehydes than with ketones and will only take place with ketones when there is a CH_3 group attached to the carbonyl group. The product of the reaction, the bisulphite compound, is crystalline and, if an aldehyde is shaken with a saturated solution of sodium bisulphite, the 'bisulphite compound' crystallizes.

It is useful at this point to consider why the anions of a strong acid will not add to the carbonyl bond. This problem has been discussed previously in Chapter 4, when we were considering the reaction of anions as electron donors in substitution reactions with the alkyl halides. It was emphasized that in the substitution reaction an important feature of such an anion was that it donated a pair of electrons to the carbon atom. The same applies to the addition of an electron donor. The only anions that will add to the carbon of the carbonyl group are those which are ready to give up their non-bonded pair of electrons. The stronger the acid, the more firmly the anion holds on to the electron pair, so that only the anions of weak acids add to the carbonyl bond. We shall consider here one further example, the addition of the hydrogen sulphide ion:

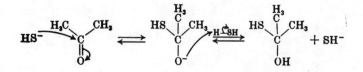

In Chapter 4 it was emphasized that the electron pair of the anion was important in initiating the displacement reaction. The discussion then turned to molecules such as ammonia which have a pair of non-bonded electrons and are capable of behaving as electron donors. The same situation applies for such addition to the carbonyl bond. Therefore, we expect ammonia and similar molecules to add to the carbon of the carbonyl bond; e.g.

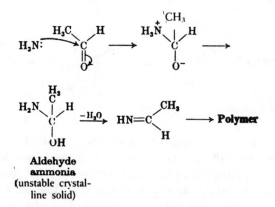

Aldehyde ammonia (unstable crystalline solid)

Under carefully controlled conditions ammonia adds directly to ethanal to yield the unstable solid 'aldehyde ammonia'; however in general the aldehyde ammonia is too unstable to be isolated and the products from the reaction between an aldehyde and ammonia are polymeric.

In some cases the structure of the polymer is known. For example, the product formed from ammonia and methanal is known as hexamethylenetetramine:

$6\,HCHO + 4\,NH_3 \longrightarrow$

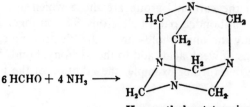

Hexamethylenetetramine

and a trimer of the product of the interaction of ethanal and ammonia has been characterized:

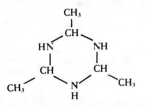

Although the reaction between carbonyl compounds and ammonia is often very complex, the reaction between carbonyl compounds and primary amines is usually more straightforward and yields an imine or Schiff base:

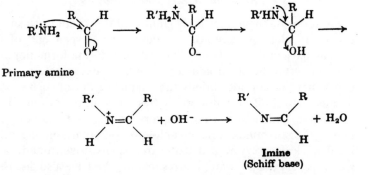

Primary amine

Imine
(Schiff base)

Reactions of this kind between a carbonyl compound and compounds containing nitrogen are both common and extremely important. Other examples include the reaction with hydrazine to yield a hydrazone:

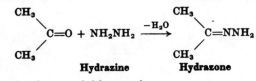

Hydrazine   Hydrazone

with hydroxylamine to yield an oxime:

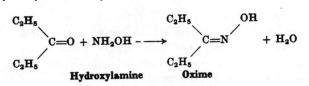

Hydroxylamine   Oxime

or with semicarbazide to yield a semicarbazone:

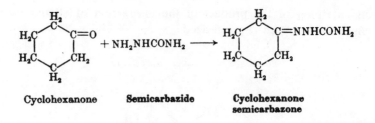

Cyclohexanone   Semicarbazide   Cyclohexanone semicarbazone

These last three reactions are important because they can be used for the identification and separation of compounds containing carbonyl groups. We shall now turn to the addition of electron acceptors to the carbonyl bond.

Addition of Electron Acceptors

From our discussion at the beginning of this chapter we should expect an electron acceptor to add to the oxygen end of the carbon–oxygen double bond. This will lead to the formation of a bond between the electron acceptor and the oxygen atom, and it is pertinent to consider briefly the characteristics of such a bond because it will differ considerably from a bond to carbon. Great stress was laid in the early chapters on the fact that the carbon–hydrogen bonds were completely non-polar; on the other hand we know very well that the oxygen–hydrogen bond, as in water, is very polar—water ionizes readily. We have also discussed the fact that the oxygen–hydrogen bond in alcohols will also ionize. This easy heterolysis of bonds between electron acceptors and oxygen means that the only important electron acceptor we need consider as adding to the carbonyl bond is the proton, because other bonds are likely to yield unstable products:

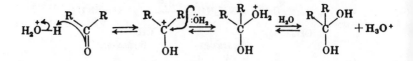

The above equation simply illustrates that the hydration of an aldehyde or ketone can be acid-catalysed in the same way as it can be base-catalysed. The fact that the hydrates are normally unstable substances existing only in solution has already been discussed. The important feature of the above equation, however, is that the first product from the addition of a proton to a carbonyl compound yields a carbocation and this species will be a much more powerful electron acceptor than the carbonyl bond itself, so that it will add electron donors which are too weak to attack the carbonyl bond in the absence of acid. An example is the reaction between a ketone and an alcohol in the presence of an acid (HX):

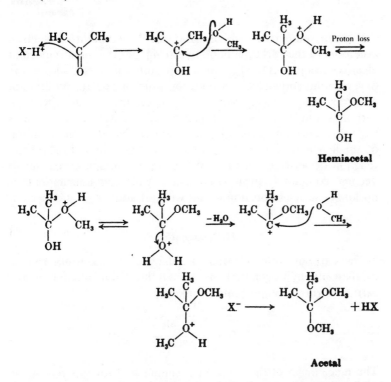

Hemiacetal

Acetal

We have used only reversible arrows for reactions involving migrations of a proton because these transformations are extremely rapid, but it is important to realize that all the steps in the above sequence are reversible. A significant feature of the sequence is the fact that the addition of the proton makes the carbonyl group a sufficiently powerful electron acceptor to add an alcohol molecule.

Thiols can be added in the same way as alcohols and the product in this case is interesting as it can be oxidized to yield sulphonal, a simple narcotic drug:

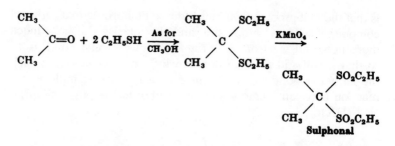

We can summarize our discussion of the addition of electron acceptors to the carbonyl bond by saying that the most important electron acceptor is the proton. The proton, of course, adds to the oxygen atom and yields a carbocation which makes the addition of electron donors easier than before. Thus, the reaction between carbonyl compounds and nitrogen-containing derivatives, such as hydrazine and semicarbazide, may be accelerated by the addition of small amounts of acid. A large excess of acid will retard the reaction because it will cause the nitrogen-containing reagent to become completely protonated, under which circumstances it is no longer an electron donor and will not add.

Hydrogenations

The carbonyl double bond in aldehydes and ketones can be catalytically hydrogenated in exactly the same manner as the carbon–carbon double bond in alkenes:

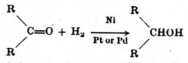

The mechanism of the reaction is similar and for this reason we will not discuss it in detail again here. The reaction is less exothermic than the hydrogenation of alkenes and thus somewhat more sluggish:

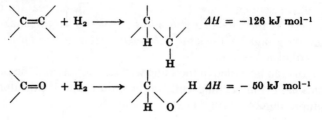

Aldehydes and ketones can also be reduced to alcohols by sodium amalgam in an acid solution. In this case the function of the metal is to provide two electrons in the reduction process:

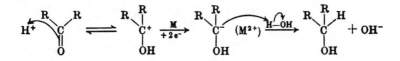

The next three reactions involve 'hydride transfer'. We are well acquainted with reactions of hydrogen atoms in which the hydrogen atom nucleus, a proton, takes part in the reaction. H^+ is something learned about at the very beginning of any study of chemistry. H^-, the hydride anion, is very much less common although it is met in compounds such as sodium hydride. Though the free existence of a hydride anion is not very common in normal systems, reactions in which a hydrogen atom transfers with two electrons are quite common, and we have already discussed one such reaction. In the previous chapter the hydroboration of alkenes was discussed and we said this reaction began by the two electrons from the carbon–carbon double bond being donated to the incomplete outer shell of the boron atom, i.e. an addition of the double bond to an electron acceptor. We then said that a hydrogen atom on the boron migrated with two electrons to complete the addition process. This reaction is an example of hydride transfer.

Returning now to consider the reduction reactions of aldehydes and ketones, we come to the Meerwein–Ponndorf reaction. In this reaction the aldehyde or ketone to be reduced reacts with aluminium isopropoxide, prepared from aluminium and isopropyl alcohol. When the ketone and aluminium isopropoxide are mixed an equilibrium is set up forming acetone and the aluminate of the alcohol derived from the reduced form of the carbonyl compound.

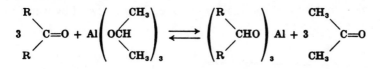

So long as the aldehyde or ketone being reduced is reasonably high-boiling the acetone can be distilled out from the reaction mixture, thus pushing the reaction to the right-hand side. The resulting aluminate can then be treated with dilute acid to yield the alcohol:

$$\left(\begin{array}{c} R \\ \diagdown \\ CHO \\ \diagup \\ R \end{array} \right)_3 Al \xrightarrow[H_2O]{H_2SO_4} \begin{array}{c} R \\ \diagdown \\ CHOH \\ \diagup \\ R \end{array}$$

The mechanism of this reaction is very similar to the hydroboration reaction described in the last chapter. The reaction is initiated by the donation of a pair of electrons from the carbonyl double bond to the incomplete outer shell of the aluminium. Subsequent steps of the reaction follow as in hydroboration:

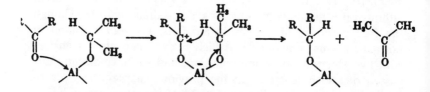

Another method of reducing aldehydes and ketones to the corresponding alcohols is to treat them with lithium aluminium hydride, $Li^+AlH_4^-$, in ether in which the lithium aluminium hydride is soluble. This reaction also involves hydride transfer:

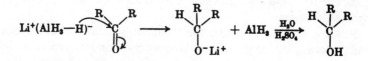

Sodium borohydride, $NaBH_4$, behaves in a very similar fashion to lithium aluminium hydride, but is a somewhat milder reagent. The mechanism of this reduction is the same as that of lithium aluminium hydride and likewise involves a hydride transfer. Aldehydes and ketones can also be reduced by diborane, B_2H_6. The trialkylborate first formed is subsequently hydrolysed by water to form the alcohol:

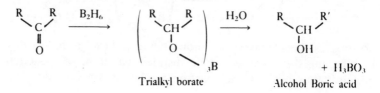

Trialkyl borate   Alcohol Boric acid

Another reaction, restricted to aldehydes which do not contain a CH_2 group adjacent to the carbonyl group, known as the Cannizzaro reaction, also involves hydride transfer. When such aldehydes are treated with alkali an equimolar mixture of the corresponding carboxylic acid salt and primary alcohol is formed:

$$2 \text{ RCHO} \xrightarrow{\text{NaOH}} \text{RCO}_2^- \text{ Na}^+ + \text{RCH}_2\text{OH}$$

This reaction occurs with methanal, HCHO, or with 2,2-dimethylpropanal, $(CH_3)_3CCHO$, but not with ethanal where the reactions are much more complex and involve the methyl group. The mechanism of the reaction involves addition of a hydroxyl anion to the carbonyl group, followed by hydride transfer from the hydrated anion to another molecule of aldehyde:

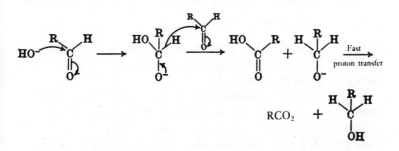

We next consider reactions in which the oxygen is completely eliminated from the organic compound to be replaced by two hydrogen atoms. The first of these, called the Wolff–Kishner reaction, involves treating the hydrazone of the aldehyde or ketone with alkali at a fairly high temperature:

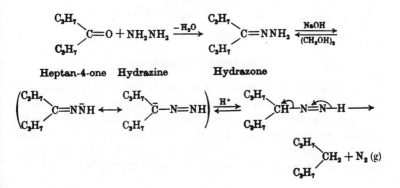

The first step in the reduction is the reversible removal of a proton from the hydrazone. This yields an anion, and we have written two structures for the anion, one with the negative charge on the nitrogen atom and the other with the negative charge on the carbon atom. We have again used the symbol of the double-headed arrow between these two structures which differ only in the positions of electrons (see Chapter 7). The implications of this symbolism will be discussed in detail in the next chapter.

Another method of replacing the oxygen in the carbonyl bond by two hydrogen atoms involves the reaction of the aldehyde or ketone with a thiol, as described above, and the treatment of the thioacetal with hydrogen and nickel:

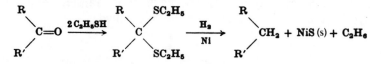

This removal of divalent sulphur from aliphatic compounds by treatment with nickel and hydrogen is a general reaction and is known as desulphurization.

A final method for directly removing oxygen from a carbonyl group and replacing it by two hydrogen atoms is by Clemmensen reduction, in which the aldehyde or ketone is treated with amalgamated zinc and hydrochloric acid. The mechanism of the Clemmensen reduction is not well understood. Probably bonding of the organic molecule to the metal surface is involved. Certainly alcohols are not intermediates as they are unaffected by amalgamated zinc and hydrochloric acid:

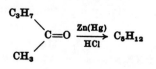

Oxidation of Aldehydes

Aldehydes can be readily oxidized to the corresponding carboxylic acids. Ketones, on the other hand, can only be oxidized by breaking the carbon–carbon chain. The latter does not occur particularly readily and will not be discussed in the present chapter. Aldehydes can be oxidized to the corresponding carboxylic acids by a wide variety of oxidizing agents; e.g. acid, alkaline or neutral permanganate. The mechanism of this reaction depends slightly on the acidity of the medium but the reaction involves, as we should expect, addition of an electron donor to the carbonyl double bond:

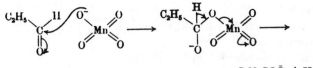

$$C_2H_5CO\bar{O} + H^+ + MnO_3^-$$

$$(2\ MnO_3^- + H_2O \longrightarrow HMnO_4^- + MnO_2 + OH^-)$$

Some aldehydes can also be oxidized by a large variety of other reagents and in some cases even by the air. Oxidation by the air involves a free-radical process for which an initiator, usually a peroxide, is required. For this reason aldehydes stored in reagent bottles are often decomposed and open bottles of benzaldehyde slowly develop crystals of benzoic acid inside them.

Nomenclature

Aldehydes are given the name of the hydrocarbon from which they are derived, followed by the suffix '-al':

$$CH_3CHO \qquad \text{Ethanal}$$
$$CH_3(CH_2)_4CHO \qquad \text{Hexanal}$$

(The semi-systematic nomenclature, by which aldehydes are named as derivatives of the corresponding carboxylic acid, for example CH_3CHO (derived from CH_3CO_2H), is called acetaldehyde, is still in general use.)

Ketones receive the suffix '-one':

$$CH_3COCH_3 \qquad \text{Propan-2-one}$$
$$CH_3CH_2CH_2CH_2COCH_3 \qquad \text{Hexan-2-one}$$
$$CH_3CH_2CH_2COCH_2CH_3 \qquad \text{Hexan-3-one}$$

(The semi-systematic name in this case uses the word 'ketone'; thus hexan-2-one becomes butyl methyl ketone and hexan-3-one becomes ethyl propyl ketone.)

Problems

1. What addition reaction, if any, would you predict between (a) ethene and (b) methanal with the following reagents?

(1) A solution of sodium cyanide to which a trace of acid has been added
(2) An aqueous solution of sodium chloride
(3) Hydrogen chloride
(4) An aqueous solution of hydroxylamine
(5) An aqueous solution of ammonia

2. What reaction, if any, would you expect between the following reagents and (a) bromoethane (b) propanone (acetone)?

<table>
<tr><td>(1) NH_3</td><td>(2) Na^+CN^-</td></tr>
<tr><td>(3) $Na^+OC_2H_5^-$</td><td>(4) H^+Cl^- in CH_3OH</td></tr>
</table>

3. Suggest a structure for a compound having the following spectral features

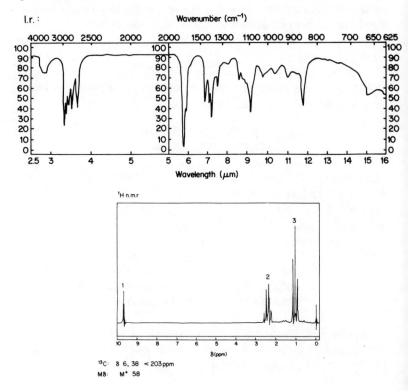

^{13}C: 8 6, 38 < 203 ppm
MS: M^+ 58

CHAPTER 9

Reactions of the OH and C=O Bonds of the Carboxyl Group

So far we have only considered the reactions of one bond at a time. We now wish to consider the carboxyl group where not only are there two 'functional groups' but they are both attached to the same carbon atom, forming what we might call a 'composite functional group'. Although the carboxyl group is apparently made up of a carbonyl and a hydroxyl function, they so modify each other that their reactions are markedly different from those of the carbonyl group in ketones or the hydroxyl group in alcohols. Compounds containing the carboxyl group are called carboxylic acids. The simplest carboxylic acid is methanoic acid (formic acid) and we can formally build up a series of straight-chain and branched-chain carboxylic acids as before (see next page).

The straight-chain carboxylic acids with eight carbon atoms or less are liquids but with chains longer than this they become low-melting solids. The lower members of this series are completely soluble in water. As with alcohols (Chapter 5), hydrogen bonding is very important in the liquid state and in solution. Carboxylic acids form dimers which persist in dilute solution in non-hydroxylic solvents (e.g. carbon tetrachloride) and in some cases even in the gas phase:

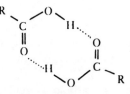

Just as ethers have lower boiling points than alcohols of considerably lower molecular weight, so methyl and ethyl esters in

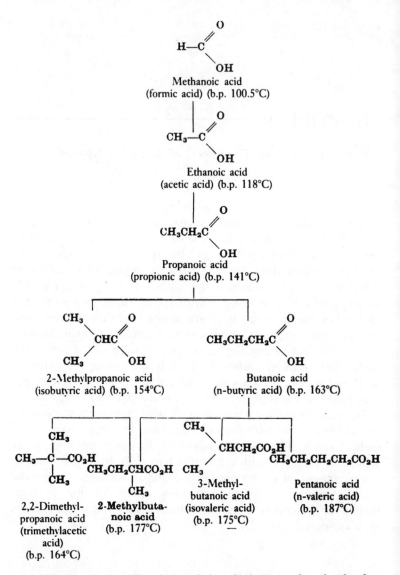

which hydrogen bonding is much less important than in the free
acid boil much lower than the carboxylic acids from which they
are derived (see Table 9.1).

The infrared spectra of carboxylic acids (e.g. see Figure 9.1)
have very characteristic broad bands ~2400 – ~3700 cm^{-1}. This
diffuse band is due to hydrogen bonding. Most of the remaining

Table 9.1 Boiling points of methyl and ethyl esters of some carboxylic acids.

Acid	B.p. (°C)	Methyl ester	B.p. (°C)	Ethyl ester	B.p. (°C)
Methanoic HCO_2H	100.5	HCO_2CH_3	32	$HCO_2C_2H_5$	54
Ethanoic CH_3CO_2H	118	$CH_3CO_2CH_3$	57	$CH_3CO_2C_2H_5$	77
Propanoic $CH_3CH_2CO_2H$	141	$CH_3CH_2CO_2CH_3$	80	$CH_3CH_2CO_2C_2H_5$	99

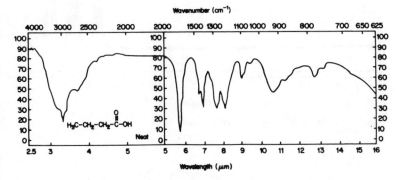

Figure 9.1 Infrared spectrum of butanoic acid.

peaks in the infrared spectrum can be attributed to groups observed in other molecules; in particular the strongest bond occurs at circa 1710–1720 cm^{-1} and is associated with the carbonyl stretching frequency.

The ^1H n.m.r. spectrum of butanoic acid (Figure 9.2) shows a characteristic pattern of absorptions attributable to a propyl group ($\overset{a}{C}H_a\overset{b}{C}H_2\overset{c}{C}H_2-$): CH_3 triplet a at δ 1.0 ppm, CH_2 multiplet b at δ 1.7 ppm and CH_2 triplet c at δ 2.39 ppm. In addition there is a singlet way downfield d at δ 10.97 ppm due to the acidic proton of the carboxyl group.

The ^{13}C n.m.r. spectrum (Figure 9.3) shows the carbonyl carbon atom right downfield (δ_C 180.69 ppm), the adjacent CH_2 group at δ_C 36.27 ppm and the remaining peaks at δ_C 18.46 and δ_C 13.70 ppm.

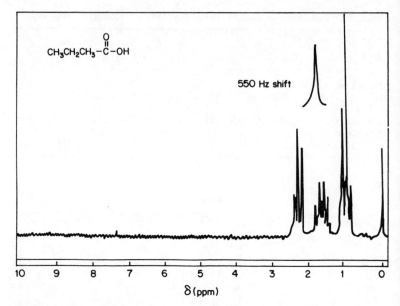

Figure 9.2 ¹H n.m.r. spectrum of butanoic acid.

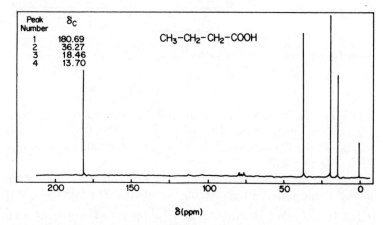

Figure 9.3 ¹³C n.m.r. spectrum of butanoic acid.

The Acidity of the O—H Bond

When discussing alcohols, we described how they could behave as very weak acids. Ethanol, with an acid dissociation constant of $10^{-18}M$, is a very much weaker acid than water, the dissociation constant of which is $2 \times 10^{-16}M$. The dissociation constant of ethanoic acid in water is $1.8 \times 10^{-5}M$:

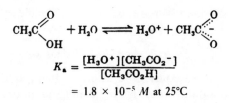

$$K_a = \frac{[H_3O^+][CH_3CO_2^-]}{[CH_3CO_2H]}$$

$$= 1.8 \times 10^{-5} \, M \text{ at } 25°C$$

This means that the molecule CH_3CO_2H is a more favourable state (of lower free energy) in water than

On the other hand, since ethanoic acid is so very much stronger an acid than water or ethanol, the ethanoate ion in water must be of lower energy (free energy) than OH^- or $C_2H_5O^-$ in the same solvent. In both these molecules the negative charge is situated exclusively on the oxygen atom. In the ethanoate anion there are two oxygen atoms to choose from and there is no means of distinguishing them. In fact the negative charge is distributed equally between the two oxygen atoms. We can either represent this as we have done in the equations above, by drawing a dotted line between the two oxygen atoms and a negative charge somewhere in between them, or we can draw two 'classical structures' with a double-headed arrow between them:

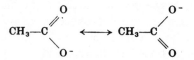

The objection to the latter method of representing the ethanoate anion is that it suggests that the negative charge is oscillating backwards and forwards between the two oxygen atoms. This is not so. In ethanoic acid the two carbon–oxygen bonds are of different lengths, approximately 124 pm for the carbonyl oxygen bond and 143 pm for the carbon–hydroxyl bond. However in the ethanoate anion the two carbon–oxygen bonds are of identical length. Although the classical structure is misleading in this respect, it is useful in that it tells us, for example, that the negative charge is situated on the oxygen atoms and not on the carbon atom as might appear from the picture we have drawn with a dotted line between the two oxygen atoms and a negative charge somewhere between. We shall find as we consider further classes of organic compounds more and more molecules that cannot be represented by a single structure.

Esterification and Hydrolysis

In Chapter 5 we considered esters as being built up formally by the elimination of water from alcohol and an organic acid. We took the reaction between ethanol and ethanoic acid to form ethyl ethanoate and water as a specific example:

$$C_2H_5OH + CH_3C\overset{O}{\underset{OH}{\diagup}} \rightleftharpoons CH_3C\overset{OC_2H_5}{\underset{O}{\diagup}} + H_2O$$

It was emphasized that this was a reversible reaction. The numerical value of the equilibrium constant, K, is approximately 4 at room temperature:

$$K = \frac{[CH_3CO_2C_2H_5][H_2O]}{[CH_3CO_2H][C_2H_5OH]}$$
$$= 4 \text{ at } 25°C$$

This equilibrium is established very slowly indeed and it is important to distinguish between this equilibrium and the dissociation equilibrium we discussed above. The reaction between ethanoic acid and water to form the hydroxonium ion and the ethanoate ion only involves proton transfer and is extremely rapid—too rapid indeed to be measured by any normal method. On the other hand, the reaction between ethanoic acid and ethanol is a complex bimolecular process and in the absence of a catalyst occurs very slowly. Ethanol is too weak an electron donor to add readily to the carbonyl double bond in ethanoic acid. However, if we have a strong acid present as a catalyst it may protonate the ethanoic acid to yield a carbocation which is sufficiently electrophilic to attack the ethanol molecule:

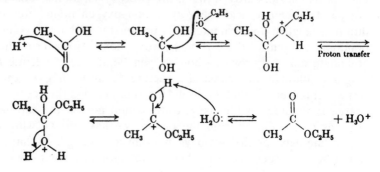

The addition of a trace of a mineral acid to a mixture of ethanol and ethanoic acid only accelerates the rate at which equilibrium between ethyl ethanoate and water on the one hand and ethanoic acid and ethanol on the other is established. In order that the reaction shall go to completion, we can either remove the water, keeping the other concentrations unchanged, or add a considerable excess of ethanol. An example of the latter method is sometimes known as the Fischer–Speier method of esterification. The carboxylic acid is treated with a vast excess of an alcohol which must be low-boiling, and dry hydrogen chloride is added as catalyst. The excess of alcohol ensures that almost all the acid is converted into the ester. The alternative way of pushing the reaction over to the ester side of the equation is to remove the water. This can be done by using concentrated sulphuric acid. The sulphuric acid serves two functions: first and most important, as a catalyst it makes the reaction go at a reasonable speed and second, as a dehydrating agent to remove the water. Another method of removing the water is by azeotropic distillation, in which a solvent is added to the reaction mixture and then distilled out again. During this distillation the solvent carries with it the water formed in the esterification.

Acid-catalysed esterification is a reversible reaction; therefore, to hydrolyse an ester, besides a mineral acid, all we require is an excess of water. This will clearly push the reaction over to the free acid and alcohol. Ethyl ethanoate, for example, is hydrolysed by heating it with dilute mineral acid. However, it is often more convenient to hydrolyse esters by base catalysis. We should expect a base such as the hydroxyl anion to add to the carbonyl bond in an ester just as it adds to the carbonyl bond of a ketone:

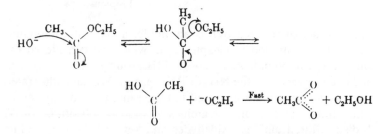

The last step in the base hydrolysis of an ester is proton transfer from the carboxylic acid molecule to the alkoxide anion. This reaction is almost irreversible, and thus base hydrolysis, unlike acid hydrolysis, of an ester is a virtually irreversible process. Base hydrolysis is in many ways a more useful reaction because the acid is now in the form of its salt which is non-volatile and insoluble in organic solvents whereas the alcohol is volatile and soluble in organic solvents; therefore the two components of the ester can be separated. It is worthwhile considering why ethyl ethanoate can be made by taking ethanoic acid, an excess of ethanol, and a trace of a mineral acid but cannot be made by taking ethanoic acid, an excess of ethanol and a trace of sodium hydroxide.

The similarity of the acid-catalysed esterification and the acid- and the base-catalysed hydrolysis of esters, with the electron-accepting and electron-donating addition reactions of the carbonyl bond in aldehydes and ketones (cf. Chapter 8), should be noted.

The Reaction of Other Electron Donors with Esters; the Formation of Amides

As the hydroxyl anion will add to the carbonyl double bond in an ester (as in ester hydrolysis), we should expect other electron donors to add in the same fashion. A particularly important example is the addition of ammonia:

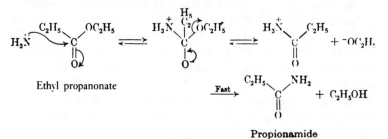

Ethyl propanonate

Propionamide

The product of this reaction is an amide, a class of compounds we briefly mentioned in Chapter 5, in which we considered the amides as compounds in which the OH group in a hydroxy acid had been replaced by the NH_2 or RNH group. We have illustrated the reaction with ammonia itself but the same reaction will occur with primary and secondary amines to yield N-substituted or N, N-disubstituted amides. Aldehydes and ketones, besides adding ammonia, also add a variety of similar derivatives, notably hydrazine and its derivatives, or hydroxylamine and its derivatives. Esters behave in the same fashion and, just as the addition of ammonia to an ester results eventually in the elimination of a molecule of alcohol, so the addition of hydrazine or of hydroxylamine also results in the elimination of a molecule of alcohol:

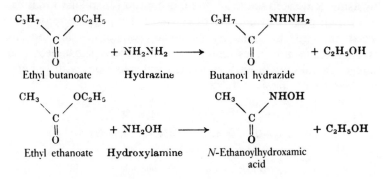

Although esters undergo these addition reactions with electron donors, note that carboxylic acids do not. In the presence of an electron donor, a carboxylic acid simply loses a proton to give the carboxylate anion and the protonated electron donor, e.g. ammonia and ethanoic acid give ammonium ethanoate. Reactions of esters with electron donors, although taking the same course as the reaction of aldehydes and ketones with electron donors, are in general much slower. This is because the alkoxyl oxygen atom to some extent diminishes the electron-withdrawing effect of the carbonyl oxygen atom. Although no complete electron transfer, as depicted in **2**, actually occurs, the carbon atom of the carboxylate group is less electropositive than it is in an aldehyde or ketone group. Thus esters react more slowly with ammonia, amines and hydrazine and will not react at all with weaker electron donors such as the bisulphite anion. In amides this effect is even stronger, and **4** makes a bigger contribution to the ground state

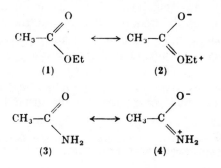

of the amide molecule. Amides do not react readily with electron-donating reagents, though they can be hydrolysed with aqueous alkali like the esters discussed above. They can also be hydrolysed by using acid catalysts; again the mechanism is similar to that for ester hydrolysis. If the ionic structure of the amide is important we should expect amides to show some acidic properties and this is, in fact, found. Thus amides are amphoteric substances and under special anhydrous conditions form salts both with mineral acids and with alkali metals:

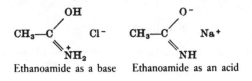

Ethanoamide as a base   Ethanoamide as an acid

Resonance Theory*

At the beginning of this chapter we found it was impossible to write a classical structure for the ethanoate anion and we suggested that one way of representing it was to draw two classical structures separated by a double-headed arrow. We were very careful to emphasize that the double-headed arrow does not mean that the negative charge is rapidly oscillating from one oxygen atom to another. The double-headed arrow is intended to indicate that the ground state of the ethanoate anion cannot be represented by either single structure but is somewhere between the two of them, i.e. with half a negative charge on each oxygen atom. In the last section of this chapter we have introduced two electronic structures for ethyl ethanoate and ethanoamide. These differ very substantially from the ethanoate ion case because whereas the two structures of the ethanoate ion are of equal energy the two structures for ethyl ethanoate (1 and 2) or for ethanoamide (3 and 4) are of very different energy. The electronic distribution in ethyl ethanoate is fairly accurately represented by structure 1 but there is a very slight migration of charge from the alkoxyl oxygen atom to the carbonyl oxygen atom. Complete transfer of charge is represented in structure 2. We therefore say that structure 2 makes a very small contribution to the ground state of ethyl

---

* This section can be omitted in a first reading. It deals with more advanced theory which is not essential to the present discussion.

ethanoate which is more accurately represented by structure **1**. When we come to the amides the transfer of charge is much more important and structure **4** makes a much bigger contribution to the ground state of ethanoamide. This is to be expected because we know oxonium ions are, in general, much less stable than ammonium ions. Let us now consider the three examples: ethyl ethanoate, ethanoamide and the ethanoate anion. The electronic distribution in ethyl ethanoate is fairly accurately represented by structure **1** although if we want to give a slightly more accurate picture we can say that structure **2** also contributes very slightly to the ground state of the molecule. With ethanoamide structure **3** does not give a very accurate picture of the electronic distribution in the ground state of the molecule although it is closer to it than structure **4**. We can in this case say that **3** is still the predominant form but that **4** makes an appreciable contribution to the ground state. Finally, we come to the ethanoate anion. In this case a single electronic structure is definitely incorrect and the true electronic distribution in the ethanoate anion can best be depicted by saying that the two forms make equal contributions to the ground state of the molecule.

This way of depicting organic structures is called 'resonance theory'. This is a most unfortunate name because it once again implies that the electrons are jumping backwards and forwards between the two structures. We can only emphasize for the third time that this is *not* the case. The term 'resonance' is derived from the quantum mechanical theory from which resonance theory is derived. According to resonance theory, the ethanoate anion is not represented by either of the two forms but, as we have indicated, is a combination of the two of them and is called 'the resonance hybrid'. Resonance theory further postulates that if a molecule can be represented as a hybrid of a number of electronic structures then the hybrid will be of lower energy than any of the constituent structures. We argued at the beginning of this chapter that ethanoic acid was a stronger acid than water or ethanol because in the ethanoate anion the negative charge is spread over two oxygen atoms and not isolated on one oxygen atom as it is in the hydroxyl anion or the ethanolate anion. This situation is described in resonance theory by saying that the ethanoate anion is a resonance hybrid of the two simple structures and is therefore of lower energy than either of the two simple structures with the negative charge isolated on the oxygen atom. Resonance theory

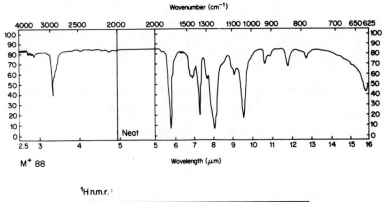

M⁺ 88

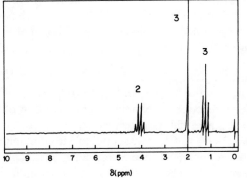

¹H n.m.r.:

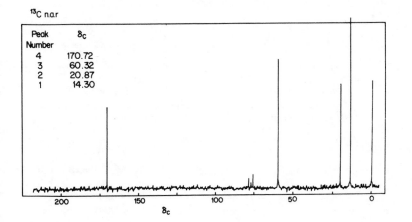

¹³C n.a.r

Peak Number	δc
4	170.72
3	60.32
2	20.87
1	14.30

CHAPTER 10 —————————————

Further Reactions of the Carboxyl Group

When considering the reactions of alcohols in Chapter 5, we discussed the reaction of an alcohol with phosphorus pentachloride to produce the alkyl chloride, phosphorus oxychloride and hydrogen chloride:

$$ROH + PCl_5 \rightarrow RCl + POCl_3 + HCl$$

The same reaction occurs with the OH portion of the carboxyl group. These reactions probably involve addition of the carbonyl oxygen to the acid chloride, which acts as an electron acceptor. Thionyl chloride, $SOCl_2$, the acid chloride of sulphurous acid, is usually a better reagent than phosphorus pentachloride because the inorganic products from the reaction—sulphur dioxide and hydrogen chloride—are both gases:

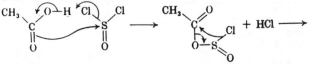

Addition of $SOCl_2$ to the C=O bond followed by the elimination of HCl

An unsymmetric anhydride. Addition of Cl to C=O followed by the elimination of SO_2.

$$CH_3C{\overset{O}{\underset{Cl}{}}} + HCl + SO_2$$

Ethanoyl chloride

Ethanoyl chloride is the first member of a general series of compounds called *acid chlorides* in which the hydroxyl group of the carboxyl composite group has been replaced by a chlorine atom. In Chapter 4 we considered how electron donors reacted with alkyl halides to give a substitution reaction, e.g.

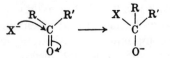

and in Chapter 8 we considered how electron donors add to the carbonyl double bond, e.g.

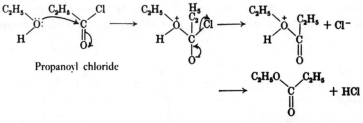

We therefore expect acid chlorides to react extremely rapidly with electron donors. Thus, even such an unreactive electron donor as ethanol reacts rapidly with propanoyl chloride:

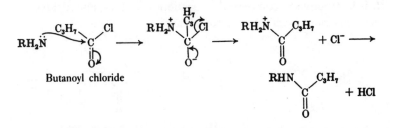

In a similar way acid chlorides react with water to regenerate the starting acid and hydrogen chloride. Amides can be prepared by the reaction of amines with acid chlorides:

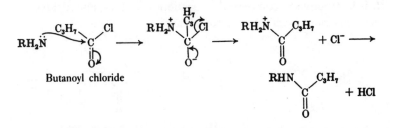

Thus acid chlorides react with the common electron donors to produce the corresponding acyl derivatives; e.g. with sodium cyanide they produce the acyl cyanides ($RCOCl + CN^- \rightarrow RCOCN + CL^-$).

A particularly interesting reaction is that in which the ethanoate anion acts as an electron donor:

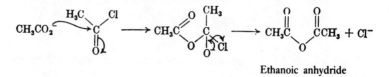

Ethanoic anhydride

Acid anhydrides, so called because they correspond to the combination of two molecules of acid by elimination of one molecule of water, are compounds of considerable importance in their own right. Ethanoic anhydride, for instance, is a commercial product used, among other things, in the manufacture of acetate rayon, a synthetic fibre. It is not made by the reaction of ethanoyl chloride and sodium ethanoate industrially, although this is a perfectly general method of preparing anhydrides. Anhydrides behave very similarly to acid chlorides in their reaction with nucleophiles. For instance, ethanoic anhydride reacts readily with amines:

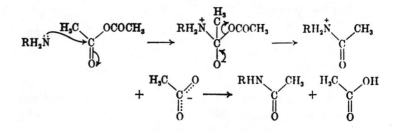

It is instructive to compare the reactions of esters, anhydrides and acid chlorides with an electron donor:

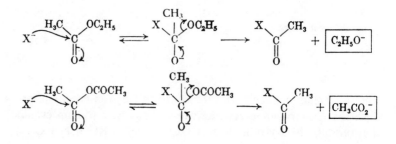

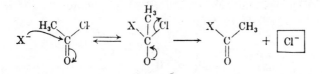

We need not consider the reaction between the electron donor and the free acid because this only involves proton transfer:

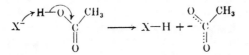

The other three reactions involve identical addition steps but in the subsequent elimination step they yield anions of increasing stability. The reaction with esters yields the ethanolate anion, the anion of an exceedingly weak acid, ethanol; the reaction with ethanoic anhydride yields the anion of an acid, but the weak acid, ethanoic acid; the reaction with ethanoyl chloride yields the chloride ion, the anion of a strong acid. Thus the reaction becomes faster down the series as a product anion becomes increasingly stable. The reaction of a secondary amine with ethyl ethanoate is extremely slow, with ethanoic anhydride it is quite rapid, but with ethanoyl chloride it can be violent and uncontrollable. Acid anhydrides and chlorides are used to prepare esters and amides. The acid-catalysed esterification of an acid is not always a practical proposition, particularly with unreactive alcohols such as tertiary alcohols, and the same argument applies to the synthesis of secondary amides.

Figure 10.1 allows the comparison of the infrared spectra of a typical carboxylic acid (note the presence of a broad OH absorption at 3500–2500 cm^{-1} superimposed on the CH absorptions at 3000 cm^{-1} and a carbonyl absorption at 1740 cm^{-1}) with the four derivatives of a carboxylic acid, all of which show the carbonyl absorption around 1750–1700 cm^{-1}, with the amide showing two absorptions for the NH_2 group at 3400 and 3500 cm^{-1}.

Reduction Reactions

Esters, anhydrides, acid chlorides and even free carboxylic acids can be reduced by lithium aluminium hydride. Like the reduction of aldehydes and ketones described in Chapter 8, this reaction involves 'hydride transfer':

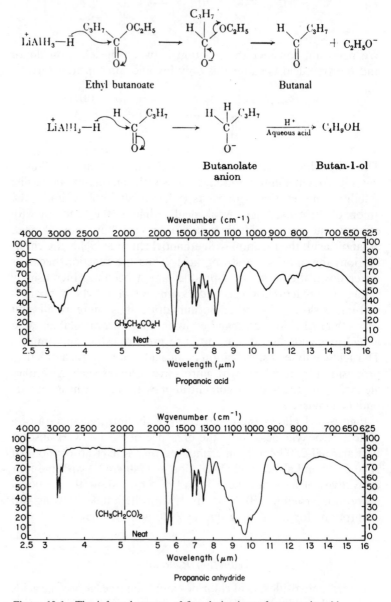

Ethyl butanoate   Butanal

Butanolate anion   Butan-1-ol

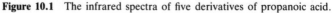

Wavenumber (cm⁻¹)

CH₃CH₂CO₂H

Neat

Propanoic acid

Wavenumber (cm⁻¹)

(CH₃CH₂CO)₂

Neat

Propanoic anhydride

Figure 10.1 The infrared spectra of five derivatives of propanoic acid.

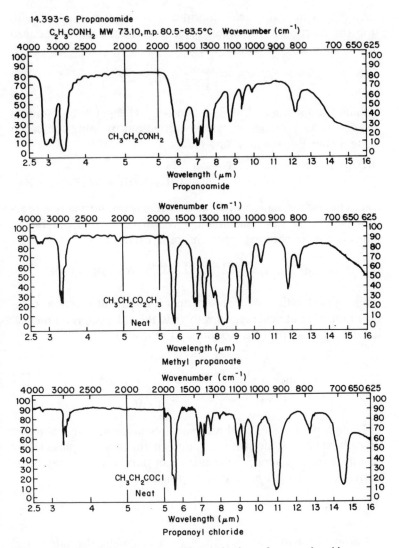

Figure 10.1 The infrared spectra of five derivatives of propanoic acid.

Notice that the first stage of the reaction yields the aldehyde which is then reduced further to the alcohol. In the reduction of esters and acids it is not practical to stop the reaction at the aldehyde stage, probably because hydride addition to the aldehyde goes more readily than hydride addition to the original ester. On the other hand, with acid chlorides it is sometimes possible to stop the reaction at the aldehyde stage. Amides can be reduced in the same way to yield amines:

$$RCONHR' \xrightarrow[\text{Ether}]{\text{LiAlH}_4} RCH_2NHR'$$

Lithium aluminium hydride is expensive; a cheaper but sometimes less satisfactory reduction can be carried out by using sodium in alcohol (the Bouveault–Blanc reaction):

$$RCO_2R' + 4\,Na + 4\,C_2H_5OH \longrightarrow$$
$$RCH_2OH + R'OH + 4\,C_2H_5O^- + 4\,Na^+$$

Catalytic hydrogenation of acids or esters is not normally a practical laboratory reaction although it can be carried out on an industrial scale with a copper chromite catalyst:

$$RCO_2R' + 2\,H_2 \xrightarrow[200°C]{\text{Copper chromite catalyst}} RCH_2OH + R'OH$$

Although catalytic reduction of esters is a difficult reaction, catalytic reduction of acid chlorides to yield the aldehyde is a practical laboratory reaction (the Rosenmund reaction):

$$RC\underset{Cl}{\overset{O}{<}} \xrightarrow{\text{H}_2 + \text{Pt catalyst}} RC\underset{H}{\overset{O}{<}} + HCl$$

Though we have cited these reactions involving the reduction of carboxylic acid derivatives, none of them occur particularly readily and, in general, the reduction of the carboxyl group or its derivatives is not to be undertaken lightly.

Nomenclature

Carboxylic acid chlorides are named by adding the suffix '-oyl chloride' to the name of the hydrocarbon from which the acid is derived.

$$CH_3COCl \qquad \text{Ethanoyl chloride (acetyl chloride)}$$
$$\underset{\underset{CH_3}{|}}{CH_3CH_2CH_2CHCOBr} \quad \text{2-Methylpentanoyl bromide}$$

Problem

1. Starting from ethanoic acid (acetic acid) how would you prepare the following compounds?

(a) Ethyl ethanoate (ethyl acetate) (b) Ethanamide (acetamide) (c) Ethanoic anhydride (acetic anhydride) (d) *N,N*-Dimethylethanamide (*N,N*-dimethylacetamide).

CHAPTER 11 ――――――――――――――

Carbon Derivatives of the Inorganic Oxyacids

The previous chapter showed how carboxylic acids formed derivatives such as esters and amides. This chapter will deal with a few analogous derivatives of inorganic acids.

Derivatives of Sulphuric Acid

In a formal way we can replace the hydrogen atoms in sulphuric acid severally by methyl groups to yield methyl hydrogen sulphate and dimethyl sulphate. These are both esters of sulphuric acid. We can also replace a complete hydroxyl group by a methyl group and this gives us methansulphonic acid; if both hydroxyl groups are replaced in this fashion we would obtain dimethyl sulphone:

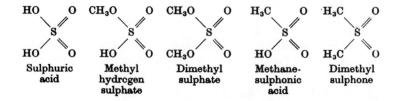

In Chapter 5 we considered esters of inorganic acids as being derived formally from the inorganic acid by replacing the OH group in the inorganic hydroxy acid by the alkoxy group of an alcohol. In Chapter 7 we showed how ethyl hydrogen sulphate can be prepared from ethene and sulphuric acid. The sulphate esters can also be prepared directly from alcohols and sulphuric acid. We have discussed at some length how the esters of carboxylic acids usually react with electron donors in such a way that the carbon–oxygen bond of the alcohol remains intact:

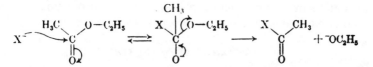

Esters of strong acids behave very differently. We could, for example, regard chloroethane as the ester of ethanol and hydrochloric acid. Chloroethane and an electron donor undergo a displacement reaction. In a similar way esters of sulphuric acid undergo displacement at a carbon atom, rather than addition to the sulphur atom:

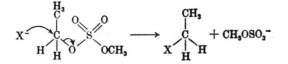

We see that if X is a hydroxyl anion and the production of the reaction is ethanol, the oxygen atom in the ethanol will be derived from the hydroxyl anion.

Organic amides of sulphuric acid are not particularly important but the amides of sulphonic acids are of great pharmaceutical interest as the so-called sulpha drugs or sulphonamides (extremely important antibiotic drugs) have the basic structure RSO_2NHR.

Derivatives of Chromic Acid

Although chromic acid esters are not readily isolated, they are believed to be intermediates in oxidation reactions involving chromic acid. Ethanol, for example, is converted into ethanal when treated with chromic acid. This reaction is believed to involve the initial formation of the chromate ester of ethanol which then reacts with water to yield the aldehyde molecule and an unstable Cr^{IV} derivative:

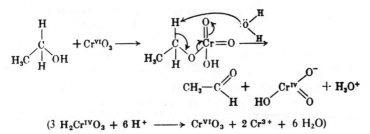

$$(3\ H_2Cr^{IV}O_3 + 6\ H^+ \longrightarrow Cr^{VI}O_3 + 2\ Cr^{3+} + 6\ H_2O)$$

Oxidation by chromic acid can often be carried out under very mild conditions and has many practical uses.

Derivatives of Nitric Acid

In Chapter 5 we described how esters of nitric acid could be considered as being derived formally from the elimination of water between nitric acid and ethanol. We also discussed how the nitrate esters were explosive substances and glyceryl trinitrate is manufactured and called 'nitroglycerine'.

The amides of nitric acid are known as *nitramines* and they and the esters can be formed by the direct action of nitric acid with the alcohol or amine:

$$RNH_2 + HO—NO_2 \longrightarrow RNH—NO_2 + H_2O$$
Nitramine

In many case sulphuric acid is used both as a catalyst and as a dehydrating agent. Nitramines, like nitrate esters, are explosive; the nitramine obtained by treating hexamethylenetetramine (see Chapter 8) with nitric acid is known as RDX, and was one of the major explosives used by the Allies during the Second World War:

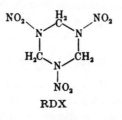

RDX

Derivatives of Nitrous Acid

Nitrite esters can be prepared directly from aqueous nitrous acid and alcohols:

$$C_2H_5O—H + HO—NO \longrightarrow C_2H_5ONO + H_2O$$
Ethyl nitrite

The amides of nitrous acid are of considerable interest and importance. Tertiary aliphatic amines having no replaceable hydrogen do not react with aqueous nitrous acid. Secondary amines form the expected nitrosamines. These are yellow oils which are less soluble in aqueous hydrochloric acid than the amine from which they are derived so that an acid solution of a secondary amine treated with sodium nitrate yields the nitrosamine as a yellow oily precipitate:

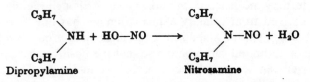

Dipropylamine   Nitrosamine

With primary aliphatic amines complications arise, for the initially formed nitrosamine undergoes rearrangement as shown in the following sequence:

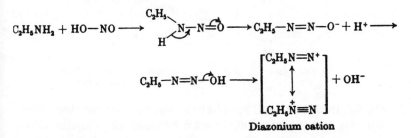

Diazonium cation

The production of these rearrangements is a diazonium hydroxide. Notice that one of the canonical forms drawn for the diazonium cation has a triple bond, i.e. six electrons shared between the two nitrogen atoms. This is the arrangement of electrons in the molecular nitrogen which we know is an extremely stable molecule. Aliphatic diazonium salts are unstable and decompose with the evolution of nitrogen, yielding a carbocation ion. In the particular case we have illustrated, the ethyl cation reacts with water to form ethanol but, contrary to what is said in many textbooks, this is not a general reaction and the carbocation may break down to yield a variety of other products, which include the alkene:

$$C_2H_5-\overset{+}{N}\equiv N \longrightarrow C_2H_5^+ + N_2(g) \overset{CH_2=CH_2+H^{\cdot}}{\underset{C_2H_5OH + H^+}{\diagup}}$$

Ethyl diazonium salt   Ethyl carbonium
(very unstable)   ion

One of the important features of these reactions is that it provides a method of distinguishing experimentally between aliphatic, primary, secondary and tertiary amines. The amine may be dissolved in dilute hydrochloric acid to which sodium nitrite solution is then added: if the amine is primary, nitrogen will be evolved very rapidly; if it is secondary, a yellow oil will be precipitated; and if it is tertiary, no reaction will take place. Although the diazonium salt derived from primary aliphatic amines has only a transient existence it can be stabilized if the diazonium group is attached to an unsaturated carbon atom so that the positive charge on the nitrogen can be spread over more of the molecule. This is what happens with the amino derivatives of benzene, which we will have a great deal more to say about later on:

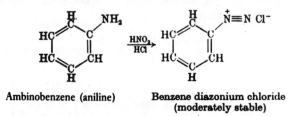

Ambinobenzene (aniline) | Benzene diazonium chloride (moderately stable)

Finally, in certain cases instead of losing nitrogen the diazonium salt may lose a proton; this is what happens when glycine ethyl ester is treated with nitrous acid and the product of the reaction is then called a *diazo compound*:

$$NH_2CH_2CO_2Et \xrightarrow{HNO_2} (\overset{+}{N}\equiv NCH_2CO_2Et) \xrightarrow{-H^+} \overset{-}{N}=\overset{+}{N}=CHCO_2Et$$
Glycine ethyl ester   Diazoacetic ester

Derivatives of Phosphoric Acid

With phosphoric acid we can clearly have three classes of ester: those in which one hydrogen atom has been replaced, those in which two hydrogen atoms have been replaced and finally the neutral triphosphate ester:

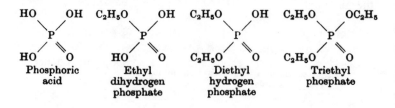

Phosphoric acid | Ethyl dihydrogen phosphate | Diethyl hydrogen phosphate | Triethyl phosphate

Organic esters of phosphoric acid are cleaved at the alkyl oxygen bond, resembling the esters of the sulphuric acids:

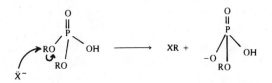

and also at the phosphorus–oxygen bond, resembling

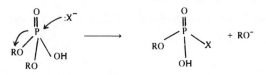

the esters of carboxylic acids.

This dual reactivity can be ascribed to the relative acidity of an alkyl phosphoric acid being less than that of a sulphuric acid, but greater than the acidity of a carboxylic acid.

We shall not discuss the reactions of phosphoric acid esters any further here, beyond briefly pointing out that the trialkyl esters can be prepared directly from phosphorus oxychloride, the acid chloride of phosphoric acid:

$$3 \ C_2H_5OH + POCl_3 \longrightarrow (C_2H_5O)_3PO + 3 \ HCl$$

The esters of phosphoric acid are of interest because they occur very widely in biological systems. In living cells phosphoric acid residues are used in a manner somewhat similar to that in which an organic chemist might be considered to use the halogens. We use a halogen in an alkyl halide as a stepping stone to convert an alcohol, for example, into a carboxylic acid. We would convert the alcohol first into the alkyl halide and then treat that with sodium cyanide and hydrolyse the resulting alkyl cyanide. Similarly, we would convert a carboxylic acid into an acid chloride before making an ester or an amide. In much the same way, living cells convert compounds into phosphoric acid derivatives as stepping stones to converting them into some other product.

Problem

1. If you were given an unknown basic substance and were told that it was aliphatic, how would you distinguish whether it was a primary, secondary or tertiary amine?

CHAPTER 12

*The Carbon–Carbon Triple Bond and the Carbon–Nitrogen
Triple Bond*

In Chapter 1 we considered a chemical bond as being formed by
the sharing of two electrons between two atoms. In Chapter 6 the
concept of a carbon–carbon double bond in which two atoms
share four electrons or two pairs of electrons was introduced. We
come now to consider a third type of bond in which two atoms
share six electrons making up three pairs of electrons. This kind
of bond is called a *triple bond*. We have already discussed two
types of double bond: the carbon–carbon double bond occurring
in ethene and the carbon–oxygen double bond occurring in aldehydes, ketones and the derivatives of carboxylic acids. We shall
similarly be concerned with two types of triple bond: the carbon–carbon triple bond occurring in acetylenes (alkynes) and the
carbon–nitrogen triple bond occurring in nitriles (cyanides). It is
worth noting that bonds between carbon atoms and other atoms
involving four pairs of electrons are not possible.

Triple bonds have a very characteristic absorption band in
their infrared spectra in the region 2100 ($RC{\equiv}C{-}H$) to 2260
($RC{\equiv}CR'$) cm^{-1}. The band is very sharp, but it may be weak.
Indeed, symmetrically substituted acetylenes show no absorption
in this region because symmetrical bond stretching results in no
change in dipole. The ${\equiv}C{-}H$ stretching gives rise to a sharp
absorption band (C–H stretch) between 3310 and 3200 cm^{-1}
(Figure 12.1).

Aliphatic nitriles have an intense $-C{\equiv}N$ stretching vibration
near 2245 cm^{-1} (Figure 12.2).

The ^{1}H n.m.r. spectra of 1,7-octadiyne and of 1,4-dicy-
anobutane (Figure 12.3) are interesting. The two methylene chains
are very similar and apart from a small chemical shift they give

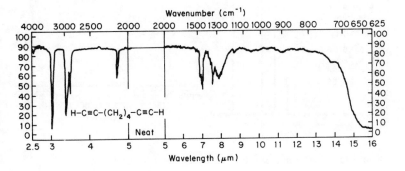

Figure 12.1 Infrared spectrum of octa-1, 7-diyne.

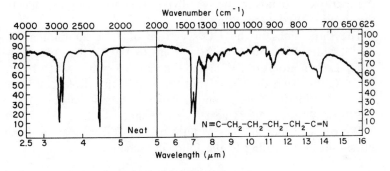

Figure 12.2 Infrared spectrum of 1,4-dicyanobutane.

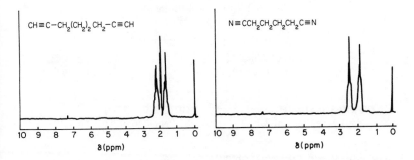

Figure 12.3 ¹H n.m.r. spectra of 1,7-octadiyne and 1,4-dicyanobutane.

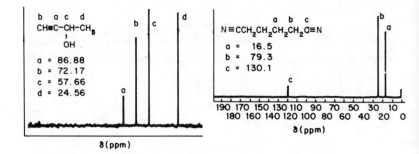

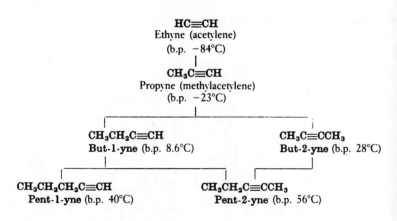

Figure 12.4 ^{13}C n.m.r. spectra of butynol and 1,4-cyanobutane.

almost identical bands. The very striking difference between the two spectra is the sharp, narrow bands occurring exactly between the methylene bands in the octadiyne. This absorption is due to the terminal hydrogen atoms.

The three lines in the ^{13}C spectrum of 1,4-cyanobutane are attributable to the nitrile carbon atoms (δ 119.3), the outer CH_2 group carbon atoms (δ 24.3ζ) and the inner carbon atom (δ 16.4). The ^{13}C spectrum of the butynol shows that the triple bond has the predominant influence and the two acetylenic carbon atoms are those showing the largest chemical shift.

We can consider alkynes as built up in a formal manner from ethyne itself according to the following scheme:

$$HC\equiv CH$$
Ethyne (acetylene)
(b.p. −84°C)

$$CH_3C\equiv CH$$
Propyne (methylacetylene)
(b.p. −23°C)

$$CH_3CH_2C\equiv CH$$
But-1-yne (b.p. 8.6°C)

$$CH_3C\equiv CCH_3$$
But-2-yne (b.p. 28°C)

$$CH_3CH_2CH_2C\equiv CH$$
Pent-1-yne (b.p. 40°C)

$$CH_3CH_2C\equiv CCH_3$$
Pent-2-yne (b.p. 56°C)

note that the ethyne molecule is linear.

Alkynes undergo addition reactions similar to those described in Chapter 7 for the addition to the carbon–carbon double bond. Again, addition can occur by four different mechanisms: electron-acceptor addition, electron-donor addition, free-radical addition and pericyclic addition (re-read Chapter 7 carefully). With hydrocarbon alkynes, the most important reaction is electron-acceptor addition. In spite of the fact that alkynes appear to be more 'unsaturated' they undergo electron-acceptor addition less readily than hydrocarbon olefins; e.g. the reaction of but-1-yne-3-ene (vinylacetylene) with one mole of bromine:

$$HC{\equiv}CCH{=}CH_2 \xrightarrow[\text{1 mole}]{Br_2} HC{\equiv}CCHBrCH_2Br$$

Apart from being slower, the addition of an electron acceptor occurs by a similar mechanism to that previously described for olefins. For instance, the addition of hydrogen fluoride to but-2-yne can be formulated as follows:

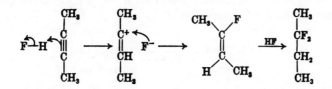

An extremely important reaction is the hydration of alkynes. We recall that the hydration of ethene occurs by the addition of sulphuric acid to the ethene molecule:

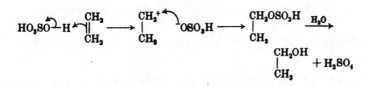

Ethyne reacts less readily with electrophiles, and hydration by sulphuric acid alone does not occur readily. However, the mercuric cation is a more powerful electron acceptor than a proton, and mercuric sulphate acts as a catalyst:

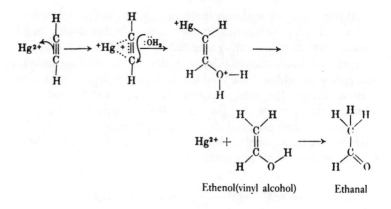

Ethenol(vinyl alcohol)   Ethanal

The first product of this reaction, ethenol, cannot be isolated. The hydrogen atom migrates spontaneously from the oxygen atom to the carbon atom to yield ethanal. The hydration of ethyne is an important industrial reaction. The ethanal so formed may be oxidized to ethanoic acid. Ethyne is thus an industrial source of ethanoic acid. The reaction is a general one and but-2-yne can be hydrated to yield butan-2-one:

$$CH_3C\equiv CCH_3 \xrightarrow[\text{H}_2\text{SO}_4]{\text{Hg}^{2+}} CH_3CH_2COCH_3$$

But-2-yne   Butan-2-one

Alkynes undergo electron-donor addition more readily than alkenes and also undergo free-radical addition under certain circumstances, but both of these reactions are outside the scope of the present book.

The triple bond in alkynes can be oxidized but the reaction occurs much less readily than with an alkene. For example:

$$CH\equiv C(CH_2)_7CH=C(CH_3)_2 \xrightarrow{\text{CrO}_3} CH\equiv C(CH_2)_7CO_2H + O=C(CH_3)_2$$

In more vigorous conditions alkynes are oxidized to two carboxylic acid molecules. If there are no CH or CH_2 groups adjacent to the triple bond, it is possible to stop the reaction at the α,β-diketone stage (A):

$$RC\equiv CR' \xrightarrow{\text{KMnO}_4} (RCOCOR') \longrightarrow RCO_2H + R'CO_2H$$

Hydrogenation of the triple bond in alkynes is an important reaction. Hydrogenation by means of gaseous hydrogen and the usual metal catalysts yields the alkane. Partial hydrogenation over a special palladium catalyst yields the *cis*-alkene whereas hydrogenation by the chemical reaction of sodium and liquid ammonia yields the *trans*-alkene:

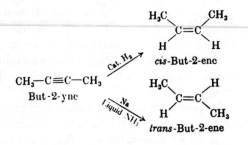

In Chapter 7 we discussed the mechanism of the catalytic reduction of alkenes and it is easy to see how a similar mechanism applied to alkynes will yield the *cis*-alkene. Chemical reduction, on the other hand, involves a two-step addition and yields the more stable *trans*-alkene.

We can regard the nitriles as being built up in a formal manner from hydrogen cyanide:

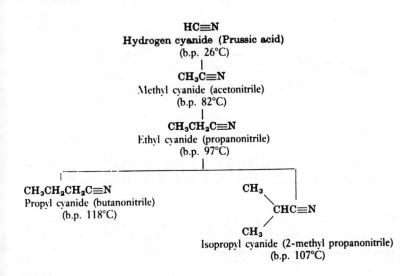

Again the hydrogen cyanide molecule is linear.

It is instructive to compare the carbon–nitrogen triple bond in cyanides with the carbon–oxygen double bond in a carbonyl compound. As oxygen is more electronegative (i.e. attracts electrons more) than carbon we argued that two types of addition reaction were to be expected with a carbonyl compound: attack by an electron donor at the carbon atom or attack by an electron acceptor at the oxygen atom. We found that the most important electron acceptor was a proton. The same is true in the carbon–nitrogen triple bond. Nitrogen is more electronegative than carbon and the cyanide triple bond can undergo addition by attack on the carbon atom by electron donors or attack on the nitrogen atom by electron acceptors.

The most important reaction of a cyanide is hydrolysis to yield first the amide and then the carboxylic acid. This reaction can be either acid-catalysed or base-catalysed. We can write the base-catalysed hydrolysis as follows:

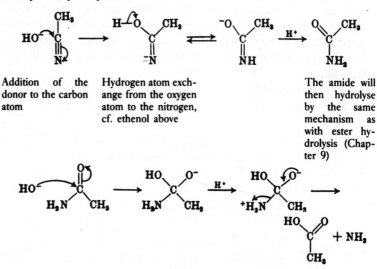

Addition of the donor to the carbon atom

Hydrogen atom exchange from the oxygen atom to the nitrogen, cf. ethenol above

The amide will then hydrolyse by the same mechanism as with ester hydrolysis (Chapter 9)

In practice acid catalysis is a more useful reaction. In this case the proton adds to the nitrogen end of the triple bond:

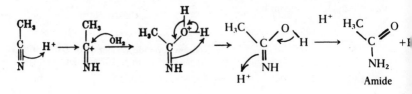

The hydrolysis of a cyanide to the corresponding carboxylic acid is of great synthetic importance. We have seen that it is possible to introduce the cyano group by replacement of a halogen. The cyano group can then be converted into the corresponding carboxylic acid. Thus we have lengthened the carbon chain.

For similar reasons the reduction of a cyano group is of synthetic importance. It can be achieved either catalytically or with lithium aluminium hydride:

$$R—C{\equiv}N \rightarrow RCH_2NH_2$$

We are used to the idea that hydrogen cyanide is a weak acid ionizing to give a proton and a cyanide anion. It seems reasonable to ask whether ethyne might not also be an acid? The answer is 'yes', although it is an extremely weak one. The sodium salt of ethyne, sodium ethynide, can be prepared by direct interaction between ethyne and molten sodium. It is prepared more conveniently by the reaction of a solution of sodium in liquid ammonia. Alternatively, sodium ethynide can be made by displacing the anion of a still weaker acid from its sodium salt; e.g. ethyne and sodamide react together to yield sodium ethynide and ammonia:

$$HC{\equiv}CH + NaNH_2 \rightarrow HC{\equiv}C^-Na^+ + NH_3$$

This reaction is also conveniently carried out in liquid ammonia solution. In Chapter 4 we considered the reaction of the cyanide anion with an alkyl halide resulting in a displacement reaction in which a carbon–carbon bond was formed and the halide ion expelled. We therefore expect that sodium ethynide would behave in the same fashion:

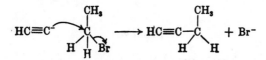

This is a general reaction and it is possible to build up a series of alkyl-substituted alkynes by this method. The compound so formed (in the present case, but-1-yne) still contains an acidic hydrogen and will therefore react with the sodamide to yield the sodium salt of but-1-yne. This can be treated with a different alkyl halide and so a disubstituted alkyne will be prepared:

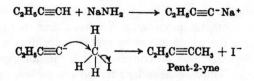

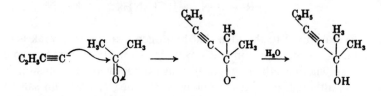

Just as a cyanide anion will add to the carbonyl bond of a ketone forming a cyanohydrin, so, in a similar reaction, an ethynide anion will add to a ketone to produce an alkynol:

The same reactions of the cyanide anions have been described in Chapters 4 and 8.

Nomenclature

Unbranched acyclic hydrocarbons having one triple bond are named by replacing the suffix '-ane' by '-yne' to the name of the corresponding saturated hydrocarbon; e.g.

$$CH_3CH_2CH_2CH_2C{\equiv}CH \qquad \text{Hex-1-yne}$$
$$CH_3CH_2CH_2C{\equiv}CCH_3 \qquad \text{Hex-2-yne}$$
$$CH{\equiv}CCH{=}CHCH{=}CH_2 \qquad \text{Hexa-1,3-dien-5-yne}$$
$$CH{\equiv}CCH_2CH_2CH{=}CH_2 \qquad \text{Hex-1-en-5-yne}$$

(When there is a choice in numbering, double bonds take the lowest number.)

Problems

1. Starting from bromoethane how would you prepare the following compounds?
   (a) Propionic acid
   (b) Hex-3-yne
   (c) Hexan-3-one
   (d) 1-Aminopropane

2. Suggest a structure which fits the following spectral information.

i.r.: 3300 (strong), 2960–2870 (strong), 2100 (medium), 1450 (weak), 1270 (weak), 620 (strong) cm^{-1}

^{1}H n.m.r.: δ 1.0, triplet, 3H; δ 1.3–1.8, multiplet, 2H; δ 1.9–2.3, multiplet, 3H

^{13}C n.m.r.: δ 13, 21, 22, 68 and 84 ppm

MS: M$^+$: 68

CHAPTER 13

Organometallic Compounds

In Chapter 4 the carbon–halogen bond was considered in the light of the first row of the periodic table. Lithium hydride is polar in the form Li^+H^- while hydrogen fluoride is polar in the direction H^+F^- and we argued that the carbon–chlorine bond would therefore be polar in the direction C^+Cl^-. We should therefore predict that in methyllithium, the carbon–lithium bond would be strongly polar in the direction $CH_3^-Li^+$. Hydrogen fluoride is a low-boiling liquid and hydrogen chloride is a gas. Lithium hydride and sodium hydride, on the other hand, are crystalline, and in the same way, methyl fluoride and methyl chloride are gases at room temperature whereas methyllithium and methylsodium are solids.

$$
\begin{array}{ccccccccc}
 & & & \mathbf{H} & & & & \mathbf{He} & \\
\mathbf{Li} & \mathbf{Be} & \mathbf{B} & \mathbf{C} & \mathbf{N} & \mathbf{O} & \mathbf{F} & \mathbf{Ne} & \\
\mathbf{Na} & & & & & & \mathbf{Cl} & \mathbf{A} & \\
\end{array}
$$

$$
\begin{pmatrix} Li^+H^- \\ Na^+H^- \end{pmatrix} \qquad \begin{pmatrix} H^+F^- \\ H^+Cl^- \end{pmatrix}
$$

The organometallic compounds of the first two groups in the periodic table are most conveniently made by direct reaction of the metal with an alkyl halide in a solvent such as ether:

$$
C_4H_9Br + 2\,Li \xrightarrow{(C_2H_5)_2O} C_4H_9Li + LiBr
$$
$$
\text{Butyllithium}
$$

$$
CH_3I + Mg \xrightarrow{(C_2H_5)_2O} CH_3MgI
$$
$$
\text{Methylmagnesium iodide}
$$
$$
\textbf{(Grignard reagent)}
$$

Alkylsodiums are so reactive that ether is an unsatisfactory solvent because the alkylsodium attacks it. Even when a hydrocarbon solvent is used the alkylsodium is so reactive that further reactions can occur. Alkylsodiums behave, as we would expect, as powerful electron donors and react with unchanged alkyl halide either by a displacement reaction or by an elimination reaction:

$$C_2H_5Br + 2 Na \longrightarrow C_2H_5^- Na^+ + Na^+Br^-$$

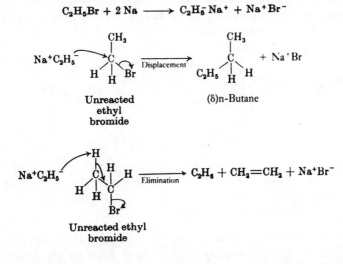

The displacement reaction which leads to the coupling of two alkyl groups is known as the *Wurtz reaction*, but this is not a reaction of practical importance.

By far the most useful of the organometallic reagants are the Grignard reagents which can be prepared directly from an alkyl bromide or iodide and magnesium in dry ether. Unlike the sodium–carbon bond, the carbon–magnesium bond is only partially polar and the Grignard reagents are soluble in ether and much more moderate in their reactions. In an alkyl halide the carbon atom attached to the halogen carries a partial positive charge. In a Grignard reagent the carbon atom attached to the metal has a partial negative charge so that, whereas the carbon atom attached to the halogen in an alkyl halide is susceptible to attack by electron donors, the carbon atom attached to the magnesium atom in a Grignard reagent is susceptible to attack by electron acceptors. The most common electron is a proton, and Grignard reagents will react rapidly with any source of a proton such as water or even an alcohol:

$$CH_3O-H \overset{\delta-}{\frown} C_2H_5 \overset{\delta+}{-} MgBr \longrightarrow C_2H_6 + CH_3OMgBr$$

$$HO-H \overset{\delta-}{\frown} C_4H_9 \overset{\delta+}{-} MgI \longrightarrow C_4H_{10} + MgIOH$$

Grignard reagents are a source of incipient carbanions (species with a negatively charged carbon atom) and will react with any compound susceptible to attack by an electron donor, e.g. they will add to a carbonyl double bond:

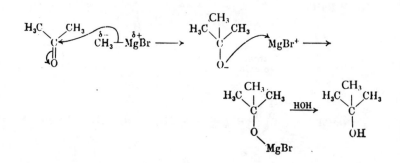

The initial adduct formed in the ether solution is decomposed by water to yield the tertiary alcohol. In practice the mechanism of the reaction is slightly more complicated than this and two moles of the Grignard reagent may be involved.

The reaction between a Grignard reagent and a ketone yields a tertiary alcohol. The reaction between a Grignard reagent and an aldehyde yields a secondary alcohol:

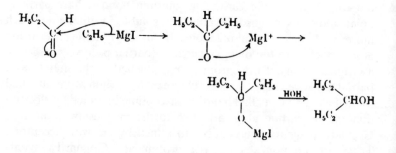

Reaction with methanal yields a primary alcohol, and this gives a method of ascending the homologous series:

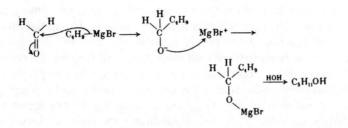

Jus as electron donors add to aldehydes and ketones so they add to the derivatives of carboxylic acids. Grignard reagents react with carboxylic esters to yield initially ketones, which react further with more Grignard reagent so that the end product of the reaction of two moles of Grignard reagent with one of ester is a tertiary alcohol:

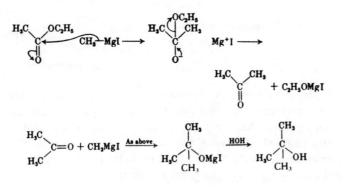

Acid chlorides behave in an analogous fashion:

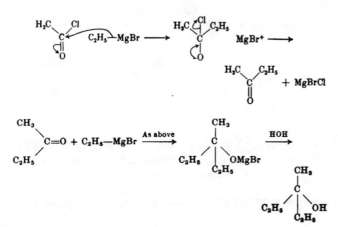

Another kind of carbonyl compound to undergo addition reaction with Grignard reagents is carbon dioxide. The initial product of this reaction is the salt of a carboxylic acid and it is possible for this to react further with more Grignard reagent. In practice, however, the second reaction is slow and in order to prepare the carboxylic acid the reaction is normally carried out by pouring an ether solution of the Grignard reagent on to powdered solid carbon dioxide. Under these conditions the reaction does not go beyond the magnesium salt of the carboxylic acid and the free carboxylic acid is liberated on addition of water. This reaction provides yet another method of ascending the homologous series:

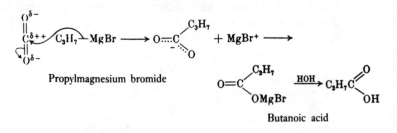

Propylmagnesium bromide   Butanoic acid

In the last chapter we discussed how electron donors would add to the carbon–nitrogen triple bond in a cyanide; therefore we should predict that Grignard reagents would add to the cyanide group in much the same way as they add to a carbonyl group. The initial products of the reaction, ketimines, are very unstable and rapidly hydrolyse to yield ketones:

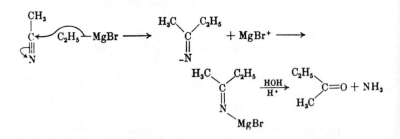

Grignard reagents can also be used for making other alkylmetal derivatives. For example, ethylmagnesium bromide reacts with lead chloride to yield tetraethyllead:

$$4\ C_2H_5MgBr + 2\ PbCl_2 \longrightarrow Pb(C_2H_5)_4 + 4\ MgClBr + Pb$$

Tetraethyllead is the most important antiknock compound added to ordinary petroleum but its use is now discouraged on environmental grounds. Commercial preparation of tetraethyllead involves the reaction of ethyl chloride with a sodium–lead alloy at moderate temperatures and pressures:

$$4\ PbNa + 4\ C_2H_5Cl \longrightarrow Pb(C_2H_5)_4 + 4\ NaCl + 3\ Pb$$

Dialkylzincs, formed from the metal and an alkyl halide, are far too reactive for ordinary synthetic work; they ignite spontaneously when exposed to the air. However, treatment of α-halo, preferably α-bromo, esters with metallic zinc dust, followed by addition of an aldehyde or ketone, results in a Grignard-type reaction. The product is a β-hydroxy ester, which can then be dehydrated to give an unsaturated ester:

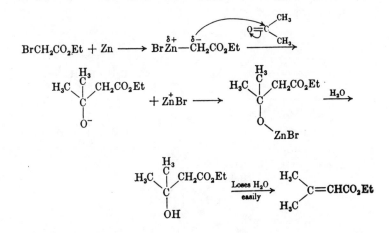

This reaction is known as the *Reformatsky reaction* and is of considerable synthetic importance in the synthesis of a class of natural products known as the terpenes.

Finally, we note that in the last chapter we discussed another class of organometallic compound, namely, the alkali metal acetylides (alkynides). Alkynes are much more acidic than saturated or alkene hydrocarbons and methylmagnesium iodide reacts with an alkyne to yield methane and the acetylenic Grignard reagent:

$$C_4H_9C\equiv C-H \quad CH_3-MgI \longrightarrow CH_4 + C_4H_9C\equiv C-MgI$$

Nomenclature

Organometallic compounds are designated by the names of the organic radicals united to the metal, followed by the name of the metal:

$$(CH_3)_2Zn \qquad \text{Dimethylzinc}$$

For compounds in which an inorganic ion is also attached to the metal, the name of this ion follows the name of the metal as in any inorganic salt:

$$C_2H_5MgBr \qquad \text{Ethylmagnesium bromide}$$

Problem

1. Starting from bromobutane how would you synthesize the following compounds?

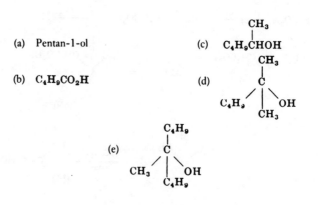

(a)   Pentan-1-ol

(b) $C_4H_9CO_2H$

(c) $C_4H_9\overset{\underset{\displaystyle |}{CH_3}}{C}HOH$

(d)

(e)

CHAPTER 14

Conjugated Dienes

In Chapter 7 we discussed the addition reaction of alkenes. We come now to consider the addition reactions of dienes in which two double bonds are separated by a single bond, as in buta-1,3-diene:

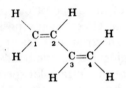

The conjugated double bonds in *trans,trans*-hexa-2,4-diene (Figure 14.1) and *cis,trans*-hexa-2,4-diene (Figure 14.2) can be distinguished in the infrared by a band at 988 cm^{-1} in the spectrum of the *trans,trans*-compound; compared with bands at 948, 982 and 1020 (weak) cm^{-1} in the spectrum of the *cis,trans*-compound.

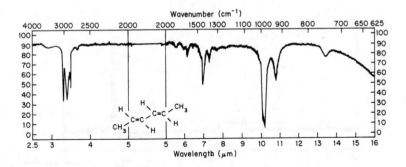

Figure 14.1 *trans,trans*-Hexa-2,4-diene.

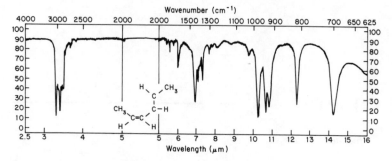

Figure 14.2 *cis,trans*-Hexa-2,4-diene.

If the conjugated polyene has other groups the assignment can be difficult.

The ^{1}H n.m.r. for both *trans,trans*- and *cis,trans*-hexa-2,4-dienes (Figures 14.3 and 14.4) are, as would be expected, very similar. The spin–spin coupling of the four vinylic hydrogen atoms can be elucidated but is out of the scope of our present discussion.

Conjugation has little effect on the position of absorption of ^{13}C in the ^{13}C n.m.r. spectrum of linear polyenes (δ 80–145).

The presence of conjugated double bonds is most readily identified by examination of the ultraviolet spectrum. As the number of conjugated double bonds increases, so both the intensity and the wavelength increases (see Chapter 2 for the u.v. spectra of unsaturated carboxylic acids).

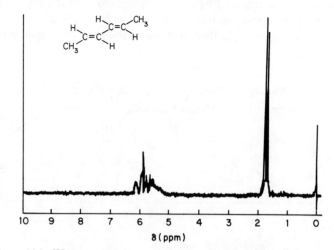

Figure 14.3 ^{1}H n.m.r. spectrum of *trans,trans*-hexa-2,4-diene.

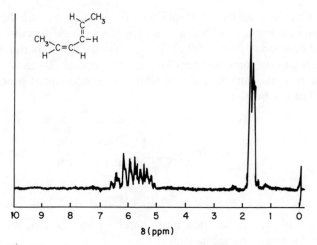

Figure 14.4 ^{1}H n.m.r. spectrum of *cis,trans*-hexa-2,4,diene.

Ultraviolet spectra for ethene and linear alkenes

Alkene	λ_{max} (nm)	ϵ
$CH_2{=}CH_2$	165	15,000
$CH_2{=}CH{-}CH{=}CH_2$	217	21,000
$CH_2{=}CH{-}CH{=}CH{-}CH{=}CH_2$	268	34,600
$CH_2{=}CH{-}(CH{=}CH)_4{-}CH{=}CH_2$	364	138,000

Infrared spectra of linear polyenes

$CH_3(CH{=}CH)_nCH_3$	Frequency (cm^{-1})
$n = 3$	1615
4	1592
5	1570
6	1561

The reactions we particularly wish to consider are additions to the double bond. As an example we choose the addition of bromine. The addition of bromine to ethene was described in Chapter 6 as a two-stage process:

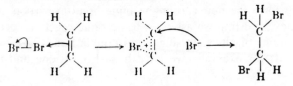

We therefore expect the addition of bromine to butadiene to occur in a similar fashion and the product of the reaction to be 1,2-dibromobut-3-ene. When we carry out the experiment we obtain two dibromobutenes. The 1,2-dibromobut-3-ene is usually present in the smaller amount while the predominant product is 1,4-dibromobut-2-ene:

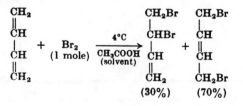

If we look at this reaction more carefully we see that we might have expected two isomers:

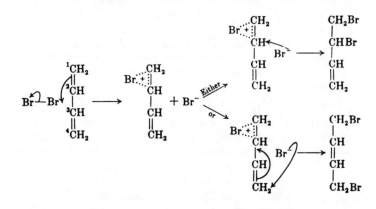

The first stage of the reaction proceeds exactly as with ethene, but there are two positions at which the bromide anion can attack. It can either attack carbon atom 2 to yield 1,2-dibromobut-3-ene or the bromide anion can attack carbon atom 4. Such an attack will force one of the pairs of the electrons in the double bond between carbons atoms 3 and 4 to rearrange to form a double bond between carbon atoms 3 and 2, leaving one of the original pairs of electrons in the double bond between carbon atoms 1 and 2 to form the new carbon–bromine bond at carbon atom 1.

The next question we ask is, will the same reaction occur with longer chains of alternate single and double bonds? The answer is 'yes'. If, for example, we have hexa-1,3,5-triene and treat it with bromine two products are obtained: 1,2-dibromohexa-3,5-diene and 1,6-dibromohexa-2,4-diene:

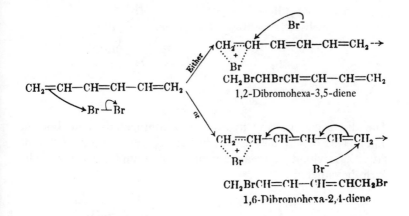

The one feature we should note about this reaction is that the 1,2-and the 1,6-dibromo compounds are formed, both of which have conjugated double bonds, i.e. with double bonds separated by one single bond. On the other hand, we get no 3,4-dibromo-hexa-1,5-diene. This suggests that a pair of conjugated double bonds is more stable than a pair of two isolated double bonds. This in turn implies that in the ground state of butadiene there must be some interaction between the two double bonds. We will not at the moment have anything further to say about the addition of electron acceptors to conjugated double bonds. We simply note that the reactions we have described for bromine are quite general for all additions of electron acceptors, e.g. for the addition of HCl, although the reaction can be complicated in this case in that it may be a reversible reaction. Not only does the addition of electron acceptors occur in a 1,4-fashion, but also of free-radicals and, when applicable, addition of electron donors can also occur in a 1,4-fashion. Conjugated double bonds will undergo all the same reactions that we described for alkene double bonds; e.g. hydrogenation usually occurs by 1,4-addition.

There is one class of addition reaction which is peculiar to conjugated dienes and this is the way they react with alkene double bonds to which a carbonyl or some similar grouping is attached. In this reaction, known as the Diels–Alder reaction, a conjugated diene reacts with a suitable alkene called a *dienophile*.

A classic example of this reaction would be that between butadiene and maleic anhydride:

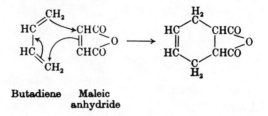

Butadiene   **Maleic**
anhydride

The diene can be any hydrocarbon conjugated diene but the
alkene (the dienophile) usually has an adjacent carbonyl or cyano
group. Acrylonitrile (cyanoethene), for example, will react with
cyclo-hexadiene:

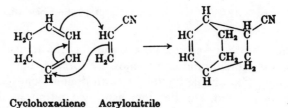

Cyclohexadiene   Acrylonitrile

The 1,4-addition and the Diels–Alder reaction are restricted to
systems of alternate double and single bonds. Thus hexa-1,3-diene
and hexa-2,4-diene both undergo 1,4-addition and will take part
in Diels–Alder additions. Hexa-1,5-diene, on the other hand,
shows none of these reactions, the two double bonds behaving
like normal completely isolated alkene bonds. We may also note
in passing that in hexa-1,2-diene (this type of compound is called
an allene) the two double bonds do not interact in the same
fashion as they do in a conjugated diene. Only in a conjugated
diene is it possible to rearrange the electron pairs down the carbon
chain when a reaction occurs:

$CH_2=CH—CH=CH—CH_2CH_3$
Hexa-1,3-diene

$CH_3CH=CH—CH=CHCH_3$
Hexa-2,4-diene

Conjugated double bonds

$CH_2=CH—CH_2—CH_2—CH=CH_2$
Hexa-1,5-diene

Isolated double bonds

$CH_2=C=CH—CH_2—CH_2—CH_3$
Hexa-1,2-diene

Cumulative double bonds

The Cyclohexatriene Problem

Let us now consider the possibility of a cyclic compound made up of alternate single and double bonds. In Chapter 6 we described how the bonds in ethene all lie in one plane subtending exactly 120° to each other. We must now add the fact that the double bond in ethene is slightly shorter (134 pm) than the single bond in ethane (154 pm). Cyclobutadiene, therefore, would not be perfectly square; nonetheless the carbon–carbon bonds would have to subtend an angle of 90° to each other. This would require substantial distortion of the alkene bond angles and, as we discussed in the first chapter, deforming bond angles introduces strain, making the molecule much less stable—so much so in this case that cyclobutadiene has not yet been isolated although there is evidence for its transient existence.

Let us now turn to cyclohexatriene. We note at once that the internal angle of a regular hexagon is 120°, i.e. exactly the same as the bond angles in ethene, so we come to the important conclusion that if cyclohexatriene exists it will be a completely planar molecule. We now have two possible structures for cyclohexatriene:

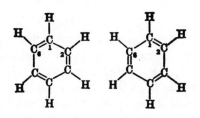

Two structures for cyclohexatriene

These structures would only be distinguishable if we could distinguish between carbon atoms 1, 2 and 6. If, for example, carbon atoms 1 and 2 were carbon-13 (^{13}C), while all the others were ^{12}C, then we could distinguish (a) which had a single bond between the two isotopic carbon atoms and (b) which had a double bond between the isotopic atoms.

Let us now compare these two structures with the structures we considered for the ethanoate anion in Chapter 9. We found that whereas in ethanoic acid the two carbon–oxygen bonds are of different lengths, in the ethanoate anion they are of identical lengths and intermediate between the lengths of the two carbon–oxygen bonds in ethanoic acid. If we slightly distort our cyclohexatriene molecule, making all the bond lengths equal and retaining the 120° angle, we obtain a molecule in the shape of a regular hexagon. We now have a situation where it is no longer possible to distinguish between the 1,2-carbon–carbon bond and the 1,6-carbon–carbon bond. We can either write our molecule as the hybrid of two electronic structures, A and B, or we can do as we did with the ethanoate ion, write a dotted line as well as a single bond C, to represent delocalization of the six electrons (one from each carbon atom in the ring):

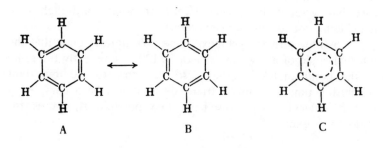

Delocalization of electrons is associated with a lowering of energy. Thus when a two or more electron structure can be drawn, the hybrid will be of lower energy than a single electronic structure. That is, cyclohexatriene will be more stable than a normal conjugated diene, just as the ethanoate anion is more stable than an ethoxide anion.

Experiments show that the molecule C_6H_6, called benzene, is extremely stable. It shows none of the reactions of a polyene. It will not react immediately with bromine. It is only hydrogenated with considerable difficulty and it is not attacked by ordinary oxidizing agents. Very powerful electron acceptors do add to benzene but instead of then adding an anion to yield a disubstituted cyclohexadiene the initial adduct eliminates a proton, thus retaining the highly conjugated benzene ring. For example, although bromine itself will not add to benzene, in the presence of a catalyst such as aluminium bromide, reaction can occur. The product of this reaction is neither 1,2- nor 1,4-dibromocyclohexadiene but bromobenzene:

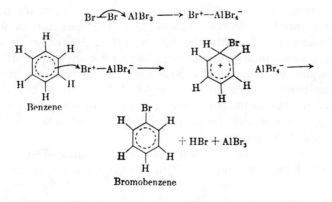

Bromobenzene

It is commonly said that benzene undergoes not addition but substitution reactions. In some ways this is a somewhat unfortunate terminology because we see in the above equations that the first step of the reaction is similar to the addition of an electron acceptor we are familiar with in the reactions of ethene. It is because of the great stability of the benzene ring that the initial adduct prefers to eject the proton ultimately to form hydrogen bromide, rather than to add the bromine anion. This addition-with-elimination reaction of benzene, resulting in substitution, is very characteristic, and benzene is considered to be the archetype of a group of cyclic molecules called *aromatic compounds* which undergo the same type of reaction.

The simpler derivatives of benzene have names as shown below:

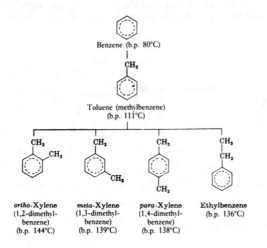

Benzene (b.p. 80°C)

Toluene (methylbenzene)
(b.p. 111°C)

ortho-Xylene
(1,2-dimethyl-
benzene)
(b.p. 144°C)

meta-Xylene
(1,3-dimethyl-
benzene)
(b.p. 139°C)

para-Xylene
(1,4-dimethyl-
benzene)
(b.p. 138°C)

Ethylbenzene
(b.p. 136°C)

Infrared spectroscopy is much used to determine the position of substituents in the benzene ring. The spectrum of toluene (Figure 14.5) illustrates the characteristic regions of i.r. absorption in a substituted benzene ring. The carbon–hydrogen stretching frequency occurs in the 3030 cm⁻¹ region and the ring carbon–carbon vibrations in the region 1650–1450 cm⁻¹. The C—H deformation of the benzene ring gives rise to absorptions between 1275 and 960 cm⁻¹ (in-plane bending) and below 900 cm⁻¹ (out-of-plane bending).

The spectra of 2-bromoethylbenzene (Figure 14.6), 3-bromo-methylbenzene (Figure 14.7) and 4-bromoethylbenzene (Figure

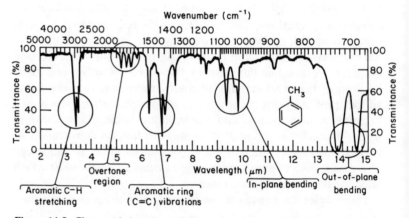

Figure 14.5 Characteristic regions of absorption of aromatic ring systems. (*Reproduced by permission of Allyn & Bacon Inc from* IR Spectroscopy, *p. 110.*)

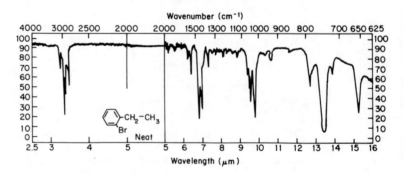

Figure 14.6 2-Bromoethylbenzene.

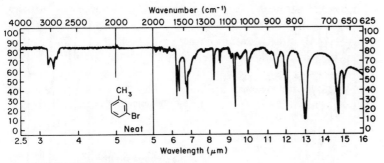

Figure 14.7 3-Bromomethylbenzene.

14.8) are shown. Of particular interest are the absorptions in the 690–890 cm^{-1} region. These are C—H 'out-of-plane' bending frequencies. *ortho*-Substituted compounds have a very strong broad band, 770–735 cm^{-1}; *meta*-substituted isomers, 810–750 cm^{-1}(sometimes there is a second *meta*-substituent band, 710–690 cm^{-1}) and the *para*-substituted isomers, 830–810 cm^{-1}. 'Overtone' bands are often too weak or too complex to be useful in structural determination: an exception is the aromatic overtone bands which occur between 6 μm (~1800 cm^{-1}) and 5 μm (~2000 cm^{-1}) (see Figure 14.9).

The ^{1}H n.m.r. absorption of the monosubstituted benzene derivatives usually appears as a singlet, but polysubstituted aromatic compounds can give very complex spectra. The ^{1}H n.m.r. spectrum of 2-bromoethylbenzene, 3-bromomethylbenzene and 4-bromoethylbenzene are shown in Figures 14.10 to 14.12. Notice

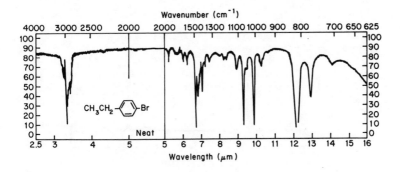

Figure 14.8 4-Bromoethylbenzene.

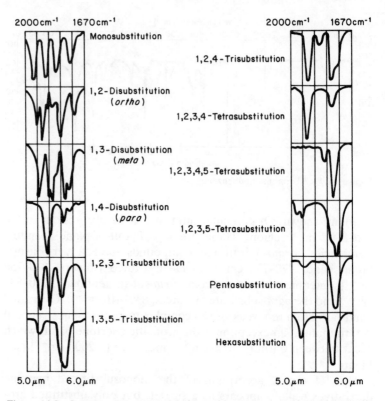

Figure 14.9 Spectra–structure correlations of benzene ring substitutions in the 5–6 μm region bands (1670–2000 cm⁻¹). (*Reproduced by permission of Allyn & Bacon Inc from* IR Spectroscopy, *page 118.*)

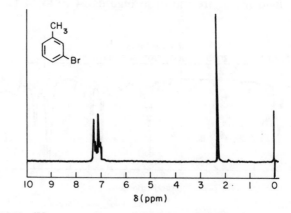

Figure 14.10 ¹H n.m.r. spectrum of 3-bromomethylbenzene.

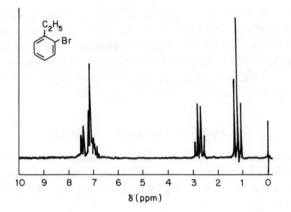

Figure 14.11 ¹H n.m.r. spectrum of 2-bromoethylbenzene.

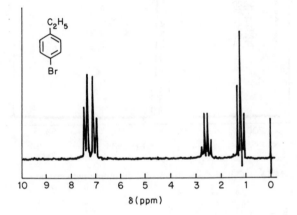

Figure 14.12 ¹H n.m.r. spectrum of 4-bromoethylbenzene.

first that the substituent methyl appears as a singlet (δ 2.35 ppm) while the ethyl groups appear as two bands (a triplet at δ ~1.20 ppm and a quartet at δ ~2.6–2.7 ppm). The hydrogen atoms attached to the benzene ring all absorb in the same region of the spectrum (δ ~7.0–7.4 ppm). The portion of the spectrum attributable to the aromatic hydrogen atoms in the 2-compound is very complex and the corresponding region in the 3-bromomethylbenzene spectrum is also complex. The aromatic region of 4-bromoethylbenzene is, in contrast, a double doublet because there are only two types of hydrogen atoms attached to the ring.

The ^{13}C decoupled spectra of the three bromotoluenes are shown in Figures 14.13 to 14.15. All three spectra have absorption bands in the δ 20–23 ppm region attributable to the methyl group. The spectrum of the *para*-isomer is very much simpler than the other spectra because there are only four types of ring carbon, whereas in the other isomers there are six.

Polycyclic Aromatic Compounds

Phenylethene ($C_6H_5CH{=}CH_2$), commonly known as *styrene*, behaves like a normal substituted ethene and undergoes the usual addition of electron acceptors; it can be readily oxidized and

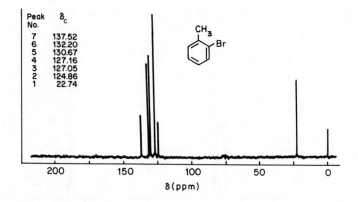

Peak No.	$δ_c$
7	137.52
6	132.20
5	130.67
4	127.16
3	127.05
2	124.86
1	22.74

Figure 14.13 ^{13}C n.m.r. spectrum of 2-bromomethylbenzene.

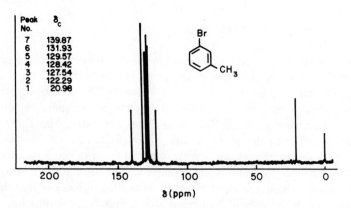

Peak No.	$δ_c$
7	139.87
6	131.93
5	129.57
4	128.42
3	127.54
2	122.29
1	20.98

Figure 14.14 ^{13}C n.m.r. spectrum of 3-bromomethylbenzene.

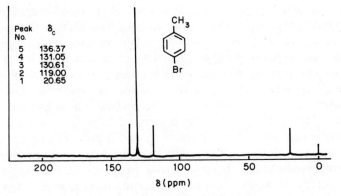

Figure 14.15 ¹³C n.m.r. spectrum of 4-bromomethylbenzene.

reduced. All these reactions occur at the side-chain double bond, the benzene nucleus remaining unaffected. Similarly 1,4-diphenyl-buta-1,3-diene (C_6H_5CH=$CHCH$=CHC_6H_5) undergoes the normal reactions of a conjugated diene. However, we can visualize a compound in which four carbon atoms are attached to a benzene ring so as to form a second benzene ring fused to the first. This compound is called *naphthalene* and it possesses most of the chemical properties of benzene.

It is possible to build up a series of condensed ring systems:

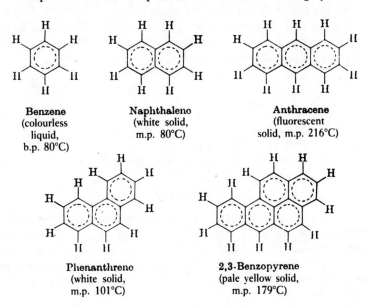

Benzene
(colourless
liquid,
b.p. 80°C)

Naphthalene
(white solid,
m.p. 80°C)

Anthracene
(fluorescent
solid, m.p. 216°C)

Phenanthrene
(white solid,
m.p. 101°C)

2,3-Benzopyrene
(pale yellow solid,
m.p. 179°C)

We have drawn two compounds containing three benzene rings: anthracene, the linear compound, and phenanthrene, the angular compound. The linear compounds become increasingly reactive as more benzene rings are fused together and even anthracene behaves in many ways more like a diene than a benzenoid aromatic compound. We have included one example of a compound containing five condensed benzene rings, 2,3-benzopyrene. This is an extremely carcinogenic compound. If painted on the skin it produces a skin cancer, and if injected it gives rise to sarcomas (tumours of connective tissue). Benzopyrene is particularly dangerous as it has been shown to be formed in the combustion processes of many organic materials and there is little doubt that it is one of the agents responsible for lung cancer.

In the next chapter we will consider the chemical properties of benzene in greater detail. The chemistry of the polycyclic aromatic compounds will not be discussed further in this book. Note, however, that although we shall discuss in detail only the chemistry of benzene, there is a large number of polycyclic aromatic compounds which have somewhat similar chemical properties.

Finally, we turn back to the question we considered at the beginning of our discussion of cyclohexatriene. Can we expect other cyclic compounds to have the great stability characteristic of the benzene molecule?

We have already considered cyclobutadiene. The next possible cyclopolyene is cyclooctatetraene. To retain the 120° bond angle of ethene the molecule is puckered. Two possible arrangements of the atoms are shown below:

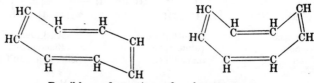

Possible conformations of cyclooctatetraene

Annulenes

It is impossible to draw two equivalent structures of cyclooctatetraene retaining the 120° bond angle without moving the carbon atoms. Two equivalent structures can be drawn if we distort the molecule in such a way that all the carbon atoms lie in a plane (it is in fact a requirement of electron delocalization that all the atoms participating in different electronic forms lie in the same plane); however, the energy required to distort cyclooctatetraene into planar form is greater than the energy that would be gained by electron delocalisation, so that cyclooctatetraene is a puckered molecule and in consequence undergoes the normal reactions of a polyene.

At first sight, cyclodecapentaene could retain the 120° angle between the carbon atoms and still be planar; however, there must then be hydrogen atoms inside the ring and these are so close together that they 'overlap'; because of this, the molecule is not, in fact, perfectly planar. The smallest compound which will retain the 120° angle and remain planar is that compound containing eighteen carbon atoms. Even here the hydrogen atoms inside the ring are so close that they probably cause slight non-planarity of the molecule.

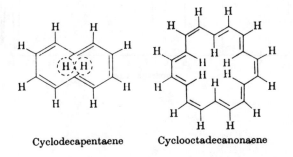

Cyclodecapentaene   Cyclooctadecanonaene

Cyclooctadecanonaene has been synethesized and proves to be a reasonably stable compound. The ¹H n.m.r. spectrum of this compound shows the presence of two types of protons: a set of twelve which absorb at about δ 8.9 ppm in the shielded region of the spectrum near where the protons of benzene absorb and a set of six protons in the de-shielded region, at about δ = −1.8 ppm. This absorption is due to the six protons inside the ring (see Chapter 2). Cyclooctadecanonaene also shows some chemical properties characteristic of benzene.

In suggesting that the unique stability of benzene can be attributed to the planar nature of the molecule and the fact that the 120° angle of the alkene double bond is not distorted, we are oversimplifying the problem. Nonetheless, these steric properties are key factors in determining the special properties of benzene.

Problems

1. What reaction would you predict between 2,3-dimethylbuta-1,3-diene and (a) 1 molar equivalent of bromine, (b) maleic anhydride and (c) ozone followed by treatment with zinc and ethanoic acid?

2. How many compounds with the molecular formula $C_{18}H_{12}$ (from benzene rings fused together) can you draw?

3. Suggest a structure for a compound having the following spectral characteristics:

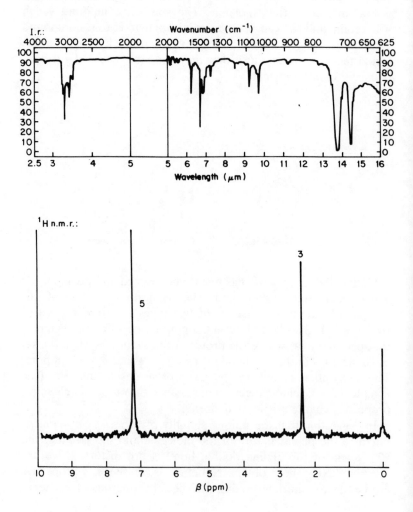

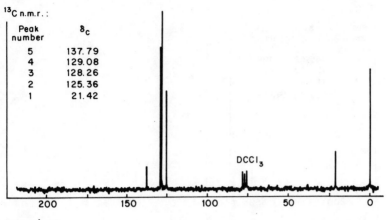

MS: M⁺ 92

4. Suggest a structure for a compound having the following spectral characteristics:

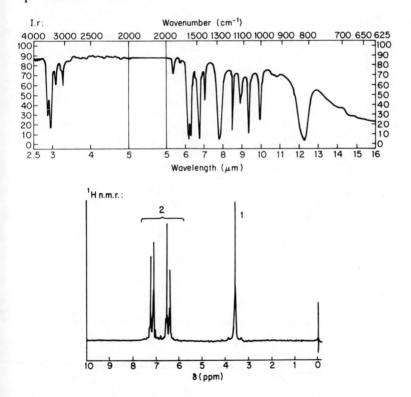

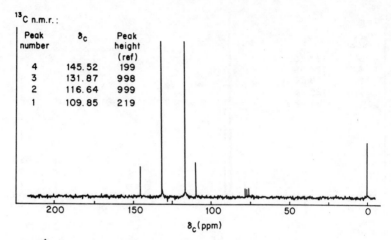

^{13}C n.m.r. :

Peak number	δ_C	Peak height (ref)
4	145.52	199
3	131.87	998
2	116.64	999
1	109.85	219

δ_C (ppm)

MS: M$^+$ a pair of peaks of equal abundance 171 and 173

CHAPTER 15

Reactions of the Aromatic Nucleus

In the last chapter we described how benzene, which does not react with bromine in the cold by itself, will react in the presence of a catalyst. The function of the catalyst is to produce an incipient bromine cation; suitable catalysts are metal halides capable of accepting an electron pair; ferric bromide is very commonly used. In the last chapter we depicted the reaction as follows:

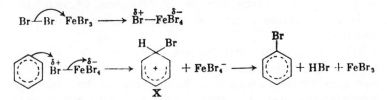

Let us take a closer look at the reaction intermediate X. If benzene is drawn as one of the two possible forms with a regular hexagonal cyclohexatriene structure* then likewise the initial adduct ion can be drawn as a hybrid of the three forms (**1a**) to (**1c**).

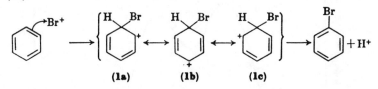

(**1a**)   (**1b**)   (**1c**)

---

* Representation of benzene is a considerable problem. The dotted circle emphasizes the fact that the electrons are delocalized and all the carbon–carbon bonds are equal. However, when discussing the reactions of benzene (and, later, when discussing the effect of substituents) in terms of resonance theory it is more useful to use the regular hexagonal hexaatriene structure, usually called a 'Kekulé structure'.

183

According to resonance theory, if a molecule can be represented by more than one electronic structure the complete resonance hybrid will be more stable than any one of the single forms. This suggests that the initial adduct of the addition of an electron acceptor to an aromatic compound is more stable than an isolated carbocation ion. We can represent this by a diagram of the kind we used in Chapter 3 when discussing the chlorination of alkanes (Figure 15.1). The adduct, often called a 'Wheland intermediate' or sometimes a 'σ complex', energy-wise lies at the bottom of a shallow valley between two peaks on the energy–reaction coordinate diagram. The first feature to notice is that the Wheland intermediate does not occur at the transition state, i.e. it is not the activated complex. There are, in fact, two transition states in this reaction designated by $\ddagger_1$ and $\ddagger_2$. (This diagram is important when considering the factors which control the rate and orientation of these reactions.) In bromination, and in the majority of other reactions that we shall discuss, $\ddagger_1$ is of higher energy than $\ddagger_2$ and the reactions are thus not readily reversible. There is one common exception to this, however: sulphonation, where the two peaks are of almost identical height; sulphonation is therefore reversible.

Before considering other reactions of benzene, which involve addition-with-elimination resulting in substitution, we shall briefly repeat our previous discussion. The first stage of these reactions proceeds in an analogous manner to the additions which occur to alkene double bonds. The difference between alkenes on the one

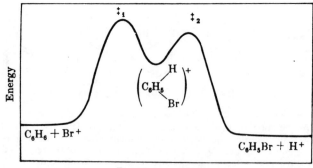

Figure 15.1 Potential energy diagram for the bromination of benzene.

hand and benzene on the other is that the initial adduct, which must always be a carbocation ion and therefore never a very stable species, stabilizes itself in the case of an alkene by adding an anion, but with benzene, because of the great stability of the aromatic nucleus, it ejects a proton. The most common electron acceptor is the proton itself. Such a reaction would therefore merely involve the interchange of two hydrogen atoms and we can only detect such a reaction by using a deuteron acid instead of a proton acid:

$$DF + BF_3 \longrightarrow D^+ - BF_4^-$$

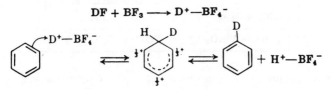

If we do this the expected replacement of a hydrogen atom by a deuterium atom takes place.

The best known of these addition-with-elimination reactions of the aromatic nucleus is nitration, but before we can discuss nitration we must take a brief look at the chemistry of nitric acid. In water nitric acid is a strong acid, i.e. it is completely dissociated, and in this sense its strength is indistinguishable from that of sulphuric acid which is also completely dissociated. However, if we take a less basic solvent such as ethanoic acid, nitric acid is then only partially dissociated, i.e. it is behaving as a weak acid whereas sulphuric acid is still completely dissociated in ethanoic acid.

$$HNO_3 + H_2O \longrightarrow H_3O^+ + NO_2^-$$

$$H_2SO_4 + H_2O \longrightarrow H_3O^+ + HSO_4^-$$

Both acids are completely dissociated in water.
(We are not concerned with the second dissociation of sulphuric acid.)

$$HNO_3 + CH_3CO_2H \xrightarrow{\longrightarrow} CH_3CO_2H_2^+ + NO_3^-$$

$$H_2SO_4 + CH_3CO_2H \longrightarrow CH_3CO_2H_2^+ + HSO_4^-$$

In ethanoic acid nitric acid is only slightly dissociated whereas sulphuric acid is still completely dissociated.

Notice that ethanoic acid, which behaves as a weak acid in water, becomes a base in nitric acid or in sulphuric acid. Nitric acid, which is a strong acid in water, becomes a weak acid in ethanoic acid. It is reasonable, therefore, to ask if nitric acid will become a base in sulphuric acid. The answer is 'yes', but the reaction is complicated by the fact that the conjugate acid is unstable and decomposes into a water molecule and a nitronium ion (NO_2^+). The water molecule produced in this reaction is immediately protonated by more sulphuric acid:

$$H_2SO_4 + HNO_3 \; \rightleftharpoons \; H_2NO_3^+ + HSO_4^-$$
$$\text{Nitric acidium}$$
$$\text{ion}$$

$$H_2NO_3^+ \quad \rightleftharpoons \quad H_2O + NO_2^+$$
$$\text{Nitronium ion}$$

$$H_2SO_4 + H_2O \; \xrightarrow{\;\;\;} \; H_3O^+ + HSO_4^-$$

$$\overline{2\,H_2SO_4 + HNO_3 \; \xrightarrow{\;\;\;} \; NO_2^+ + H_3O^+ + 2\,HSO_4^-}$$

In a sulphuric acid solution, nitric acid is converted largely into the nitronium ion. Notice that the nitronium ion can only exist in a medium in which there is no free water. The nitronium ion would react instantly with water to regenerate nitric acid. Further discussion of the properties of nitric acid are clearly not our concern at present. The important point is that in a solution of nitric acid in sulphuric acid we have a high concentration of this very reactive nitronium ion. This ion is an extremely powerful electron acceptor and we would expect it to receive an electron pair from the benzene molecule and form the Wheland intermediate. The ultimate product of this reaction is called nitrobenzene:

$$HNO_3 + 2\,H_2SO_4 \rightleftharpoons NO_2^+ + H_3O^+ + 2\,HSO_4^-$$

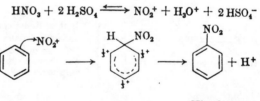

Nitrobenzene

The nitronium ion will add to alkene double bonds but in the presence of such strong electron acceptors most alkenes tend to undergo complex polymerization (cf. Chapter 7). In sulphuric acid the main electron acceptor is, of course, a proton and we have already discussed the addition of a proton to benzene. In oleum, i.e. sulphuric acid containing sulphur trioxide, sulphur trioxide can behave as a very powerful electron acceptor:

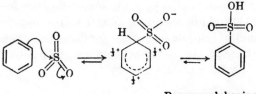

Benzenesulphonic acid

This reaction, as we have already mentioned, is reversible, unlike most of the other reactions.

Alkyl halides react with aluminium chloride to form an ionic complex in which the carbon atom originally attached to the halogen carries a positive charge. This then is capable of acting as an electron acceptor, taking a pair of electrons from the benzene nucleus:

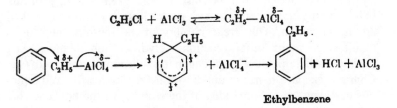

Ethylbenzene

This reaction is called the *Friedel–Crafts reaction* and is of considerable practical importance. Notice that the aluminium chloride is regenerated at the end of the sequence and is therefore only required in catalytic amounts. If this reaction occurs with an alkyl halide, it is reasonable to ask whether it will occur with an acid chloride. The answer is 'yes'. In this reaction the ketone produced forms a complex with the aluminium chloride and for this reason molecular quantities of aluminium chloride are required:

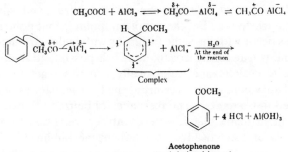

Acetophenone
(methyl phenyl ketone)

Contrary to the suggestion in many textbooks, the free acylium ion (CH_3CO^+) will not react with benzene. It is only the complex with aluminium chloride or some similar metal halide which reacts. We can see that in the acylium ion the positive charge is not sited solely on the carbon atom:

$$CH_3\overset{+}{C}{=}O \longleftrightarrow CH_3C{\equiv}O^+$$

This makes the acylium ion more stable (less reactive) than a carbocation ion and this *acylation*, as it is called, proceeds less rapidly than the alkylation described above.

These so-called 'substitution reactions' we have been discussing are of great synthetic importance. Their basic feature is that the benzene nucleus acts as an electron donor in the same way as a hydrocarbon alkene, but because of the great stability of the benzene ring the initial adduct cation ejects a proton instead of adding an anion. This great stability of the benzene nucleus is manifest in many other reactions. Hydrocarbon alkenes react rapidly with potassium permanganate solution in the cold (see Chapter 7); pure benzene undergoes no reaction with cold aqueous permanganate. Some idea of the stability of the benzene ring can be deduced from the experimental observation that methylbenzene (toluene) can be heated with a boiling solution of aqueous potassium permanganate to yield benzoic acid:

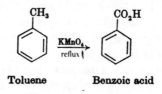

Toluene   **Benzoic acid**

Under forcing conditions benzene can be oxidized, and with ozone it yields a triozonide, but the important point is that none of these reactions occur with the same facility as with hydrocarbon alkenes. The same is true for hydrogenation. It is possible to hydrogenate benzene to cyclohexane, catalytically, but this reaction does not proceed anywhere near as readily as the hydrogenations of alkenes described in Chapter 7. The resistance of benzene to hydrogen is important in many reactions, e.g. nitrobenzene can be reduced chemically or catalytically to aminobenzene (aniline):

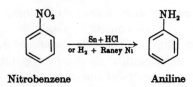

Nitrobenzene   Aniline

The difference in reactivity of the benzene nucleus and the alkene double bond is best illustrated by the reactions of phenylethene (styrene):

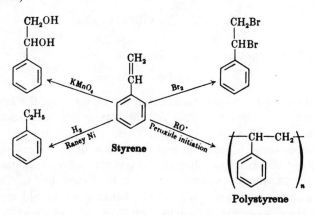

Polystyrene

Notice that styrene undergoes reactions of a hydrocarbon alkene described in Chapter 7, leaving the benzene nucleus unaffected.

The reaction of the benzene nucleus with free radicals and halogen atoms is slow; thus if toluene is treated with chlorine or bromine in the presence of ultraviolet light, reaction occurs only in the side chain:

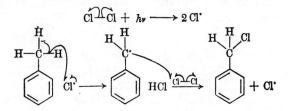

This is identical with the chain-chlorination process we described in Chapter 3. In the absence of a side chain, or when the side chain has been completely chlorinated, halogen atoms will attack the ring, and benzene with an excess of chlorine in the presence of ultraviolet or sunlight yields 1,2,3,4,5,6-hexachlorocyclohexane. This compound is sometimes most misleadingly called 'benzene hexachloride'. It will be apparent that there are a number of stereoisomers of this compound. One of them is a very powerful insecticide (lindane):

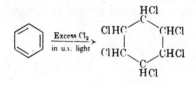

Derivatives of Benzene

So far we have been considering the reactions of the benzene nucleus. We must now consider the way in which the aromatic nucleus modifies the reactivity of substituents attached to it. Bromobenzene is called an aryl bromide as distinct from an alkyl bromide discussed in Chapter 4. Aryl halides will not undergo the displacement reactions of alkyl halides described in Chapter 4. If an electron donor is to attack benzene we should expect it to add to the aromatic nucleus. However, we already know from Chapter 7 that hydroxyl anions do not add to hydrocarbon alkenes and still less will they add to the stable aromatic nucleus. Thus, for laboratory purposes, a halogen atom attached to an aromatic hydrocarbon nucleus cannot be displaced by an electron donor. In an industrial plant, where much higher temperatures and pressures are possible, this generalization is no longer true and chlorobenzene is converted into hydroxybenzene (phenol) as an industrial process:

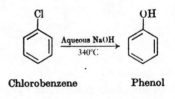

Chlorobenzene   Phenol

Although the halogens cannot be replaced by ordinary electron donors under ordinary laboratory conditions, they can be replaced by metals; bromobenzene reacts with magnesium to form phenylmagnesium bromide, i.e. the corresponding Grignard reagent, although the reaction occurs less readily than it does with an alkyl bromide:

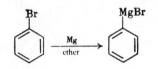

Bromobenzene   **Phenylmagnesium bromide**

Phenylmagnesium bromide will take part in all the reactions described in Chapter 12.

A hydroxy group attached to an aromatic nucleus is greatly modified in its reactions. We argued in Chapter 9 that ethanoic acid was more acidic than ethanol because the ethanoate anion could be stabilized through resonance. It is possible to write four electronic structures for the phenoxide anion (**2a** to **2d**).

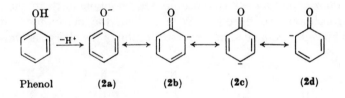

Phenol   **(2a)**   **(2b)**   **(2c)**   **(2d)**

Three of these structures have a negative charge on a carbon atom and will thus be of much higher energy than the first structure in which the negative charge is on the oxygen atom. Unlike the two structures in the ethanoate ion, which are both equally stable, the structures **2b**, **2c** and **2d**, being appreciably less stable than **2a**, will only make a very small contribution to the overall hybrid structure of the phenate anion. Nonetheless this is sufficient to make phenol very much more acidic than ethanol. Phenol behaves as a very weak acid in water (Table 15.1).

Note that if we add a proton to the nitrogen of aniline (aminobenzene) the positive charge cannot be distributed about the ring in the same way as the negative charge can be in phenol, and aniline is in fact a weaker (not a stronger) base than ethylamine:

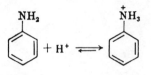

Aniline

Table 15.1 Values of K_a at 25°C in water.

Compound	K_a M
Ethanol, C_2H_5OH	$\sim 10^{-18}$
Phenol, C_5H_5OH	1.3×10^{-10}
Ethanoic acid, CH_3CO_2H	1.8×10^{-5}

However, there is a derivative of aniline in which the stabilizing effect of the benzene ring is extremely important. In Chapter 11 we discussed the reaction of aliphatic primary amines with nitrous acid and we saw that the initial product of the reaction was a diazonium salt which decomposed spontaneously to yield a carbocation and nitrogen. With aniline, the diazonium salt is stabilized to some extent by delocalization of the charge as shown:

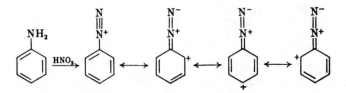

Aromatic diazonium salts can be prepared in cold aqueous solution. Crystalline diazonium salts can be isolated but they are usually very unstable. A proper discussion of the chemistry of aromatic diazonium salts is outside the scope of this introductory book. We have already inferred that nitrogen is very readily lost from these compounds and the formulae below indicate how the elimination of nitrogen can be turned into a useful preparative reaction:

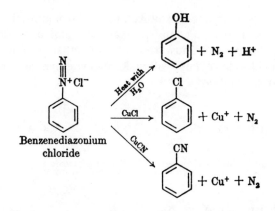

Benzenediazonium
chloride

Heating an aqueous solution of benzenediazonium chloride yields a complex tarry mixture, of which phenol is the principal component. On treatment with cuprous salts, benzenediazonium salts lose nitrogen and give good yields of the chlorobenzene (using cuprous chloride) or benzonitrile (using cuprous cyanide).

Apart from the ease with which a diazonium salt will lose nitrogen, the diazonium cation is clearly an electron acceptor and as such it will add to alkene double bonds, although this is a messy reaction; although it will not undergo an addition-with-elimination reaction with benzene it will do so with the stronger electron donor such as the phenoxide anion:

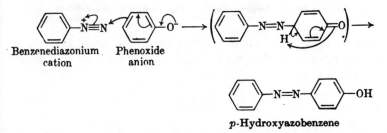

Benzenediazonium   Phenoxide
cation   anion

p-**Hydroxyazobenzene**

This coupling reaction, as it is called, is of immense importance. The product, *p*-hydroxyazobenzene, is a bright orange substance and is the simplest of the *azo dyes*. Dyes made by coupling diazonium salts with phenols, called azo dyes, are the largest single class of synthetic coloured materials used in the dyestuff industry.

Although the presence of the adjacent aromatic nucleus modifies the reactions of phenol by making it more acidic and modifies the reactions of aniline by stabilizing the diazonium salt, the main reactions of these two compounds are the same as those of their aliphatic analogues. Thus, treated with ethanoic anhydride or ethanoyl chloride, phenol will yield phenyl ethanoate and aniline will yield acetanilide:

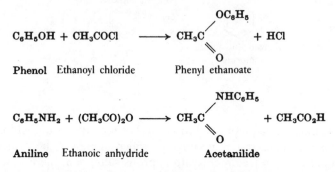

Phenol	Ethanoyl chloride	Phenyl ethanoate

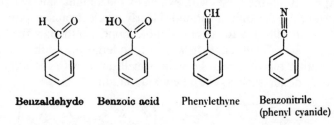

Aniline	Ethanoic anhydride	**Acetanilide**

Similarly, benzoic acid, benzaldehyde, benzonitrile and phenylethyne all undergo the reactions described in Chapters 8, 9 and 11:

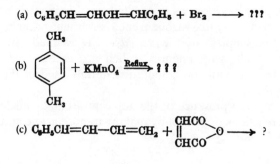

Benzaldehyde   **Benzoic acid**   Phenylethyne   Benzonitrile (phenyl cyanide)

Some books attempt to stress differences in the reactions of these compounds and their aliphatic analogues, but the important point is their similarity.

Problems

1. Predict the outcome of the following reactions:

    (a) $C_6H_5CH{=}CHCH{=}CHC_6H_5 + Br_2 \longrightarrow$???

    (b) [ring structure with CH₃ groups] $+ KMnO_4 \xrightarrow{\text{Reflux}}$???

    (c) $C_6H_5CH{=}CH{-}CH{=}CH_2 + \begin{smallmatrix}CHCO\\CHCO\end{smallmatrix}{>}O \longrightarrow$?

2. Suggest steps for carrying out the following transformations:

(a) C_6H_5Br ⟶ $C_6H_5CO_2H$

(b) C_6H_6 ⟶ $C_6H_5NH_2$

(c) $C_6H_5COCH_3$ ⟶

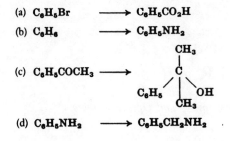

(d) $C_6H_5NH_2$ ⟶ $C_6H_5CH_2NH_2$.

CHAPTER 16

Molecular Asymmetry and Optical Activity

In the first chapter, during a discussion on the tetrahedral arrangement of bonds around a tetravalent carbon atom, we drew attention to the fact that there are two forms of lactic acid (**1a** and **1b**) (2-hydroxypropanoic acid, the systematic name for this acid, is rarely used).

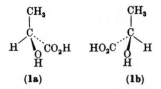

(**1a**)   (**1b**)

Let us look at this problem again. These two molecules are identical in every feature except in the disposition of the various groups in space. One is the mirror image of the other and we compared the two forms to a pair of gloves. Another analogy would be to compare them to a left-hand screw and a right-hand screw. In fact a molecule with a helical shape is known and both the left-hand helix and the right-hand helix have been separately obtained.

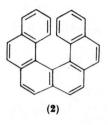

(**2**)

Compound **2**, correctly a phenanthrophenanthrene, was appropriately named 'hexahelicene'. In the formula shown above, the hydrogen atoms attached to the benzene rings have been omitted and the two terminal benzene rings have deliberately been pulled apart in the diagram. In **3a** and **3b** we attempt to represent the two forms of the molecule in three dimensions. Compounds with up to twelve benzene rings joined in this way have been made and separated into enantiomers.

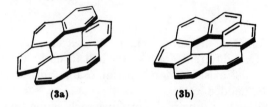

(3a)   **(3b)**

Other examples of molecules which exist in non-superimposable mirror images are 2,2′-dinitro-6,6′-diphenic acid and cyclopropane-1,2-dicarboxylic acid. In the former example (**4**) the bulky carboxyl and nitro groups prevent rotation about the bond joining the two benzene rings and thus it is possible to have two mirror images which are not superimposable. *trans*-Cyclopropane-1,2-dicarboxylic acid can exist in two non-superimposable mirror images (**5a**). The *cis*-acid, on the other hand, only has one form; its mirror image is superimposable on itself (**5b**).

Molecules such as lactic acid, hexahelicene, 2,2′-dinitro-6,6′-diphenic acid and *trans*-cyclopropane-1,2-dicarboxylic acid which can exist as two non-identical mirror images are called *chiral molecules*, and the two forms are known as *enantiomers*. Chiral molecules do not necessarily contain carbon, but chirality is par-

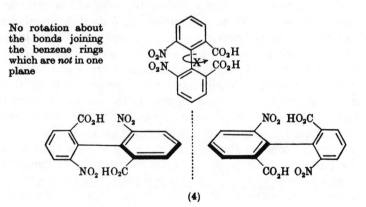

(4)

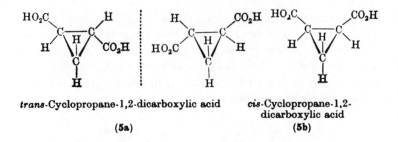

trans-Cyclopropane-1,2-dicarboxylic acid

(5a)

cis-Cyclopropane-1,2-
dicarboxylic acid

(5b)

ticularly common in carbon chemistry because a tetrahedron which has four different groups attached to each corner of it is asymmetric and exists in non-superimposable mirror images (6) (chirality ≡ handedness).

It is obvious that a pair of molecules such as the two forms of lactic acid must have identical chemical properties. Likewise, their boiling point, melting point, solubility and similar physical properties are identical. How then do we know that there are two forms and how can we distinguish them? Before we can discuss this we must briefly remind ourselves of the nature of light and in particular of polarized light.

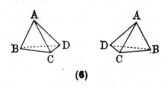

(6)

Polarized Light

There is ample physical evidence to make us associate light with wave motion. Wave motion may be caused by longitudinal vibrations or transverse vibrations. In a longitudinal vibration the vibrations are in the plane of the direction of propagation. A familiar example of longitudinal vibrations are sound waves. In transverse vibration the vibrations are perpendicular to the direction of propagation, such as ripples on the surface of water. In Figure 16.1 each vertical arrow represents a vector which maintains a fixed position and direction but varies continuously in magnitude from +1 through zero to −1 and back again during the passage of the wave. In an ordinary wave of light these transverse vectors are completely symmetrical to the direction of propagation (see Figure 16.2). Any single vector may be regarded

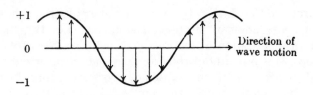

Figure 16.1 Transverse vibration.

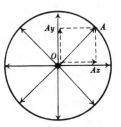

Figure 16.2 Transverse vectors in an ordinary light beam.

as the resultant of two components vibrating at right angles to each other, for example, vector **OA** may be regarded as the resultant of two vectors **OAy** and **OAx**. Thus an ordinary ray of light may be divided into two separate rays in which the transverse vibrations are all in one direction, those of one ray being perpendicular to those of the other. The light wave in which all the transverse vibrations are in one direction is said to be *plane polarized*. There are many ways in which ordinary light may be decomposed into two plane polarized rays, the most common occurring when light passes from one medium into another. A parallel beam of ordinary light striking the surface of some transparent material at the critical angle is divided into two components, a reflected ray and a transmitted ray. These rays are both plane polarized at right angles to each other. Another way in which a beam of ordinary light may be divided into two beams of plane polarized light is by passage through certain crystals.

Optical Activity

If we shine a beam of plane polarized light through a solution of lactic acid containing only one form (i.e. one enantiomer, say **1a**), we find the plane of polarization of the emergent light has been rotated to the right. If a beam of plane polarized light passes through a solution of the other enantiomer (i.e. **1b**), then the plane of polarization of the emergent light has been rotated to the left. This property of rotating the plane of polarized light is called *optical activity* and substances possessing this ability are said to be optically active. The number of degrees through which the plane of polarized light is rotated is called the *optical rotation* of the optically active substance and is given the symbol α. The direction of rotation of the plane of polarization is designated by a '+' sign for rotation to the right and by a '−' sign for rotation to the left. So **1a** is (+)-lactic acid and **1b** is (−)-lactic acid. The words *dextro* and *laevo* are sometimes used as prefixes to the names of compounds to denote the sign of rotation, i.e. (+)-lactic acid is a shorter way of writing *dextro*-lactic acid. A single enantiomer will rotate a plane of polarized light whether as a pure liquid in solution or as a vapour.

The extent of rotation of the plane of polarized light depends on the number of optically active molecules through which the light passes. Hence, the observed rotation is directly proportional to the length of the light path through the optically active material, and also to the concentration in weight per unit volume of the optically active substance in solution:

$$[\alpha] = \frac{\theta \times 100}{l \times c}$$

where $[\alpha]$ = specific rotation of the substance
   θ = the observed rotation
   l = length of the light path through the solution in decimetres
   c = concentration of the optically active substance in grams per cm^3 of solution

The specific rotation of a substance depends on the wavelength of the light, the nature of the solvent, the concentration and the temperature. It is expressed in degrees. Sodium light is usually employed and thus a specific rotation at 25° is normally reported as $[\alpha]_D^{25}$.

Optical Isomerism

The two enantiomers of lactic acid are suitably called *optical isomers*. If we prepared lactic acid from the cyanohydrin, formed when hydrogen cyanide adds to ethanal, we obtain a substance which shows no optical activity:

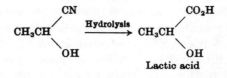

This is hardly surprising for if we look at the reaction in which the asymmetric molecule was formed, i.e. the addition of hydrogen cyanide to ethanal, we can clearly see that the cyanide ion may add to either side of the carbonyl group:

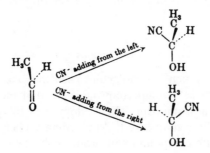

In the equation we have attempted to represent ethanal with the methyl group projecting out of the plane of the paper and the hydrogen atom behind it. It is then possible for the cyanide anion to add either from the left or from the right to yield the two possible enantiomers of the cyanohydrin. Clearly addition at both sides of the ethanal molecule is equally probably and therefore we obtain exactly equal amounts of the two enantiomers of the cyanohydrin. The subsequent hydrolysis of the cyanohydrin to yield lactic acid does not affect the asymmetric carbon atom. This is a completely general observation and any ordinary synthetic reaction in a laboratory will always produce exactly equal amounts of two enantiomers. Where then do we obtain samples containing only one enantiomer and showing optical activity? The answer is from living organisms.

Lactic acid can be extracted from muscle tissue with water. On purification it yields hygroscopic crystals, melting at 26°C and showing optical activity $[\alpha]_D^{25} + 4°$. Lactic acid can also be obtained from the action of certain microorganisms on milk sugar (lactose). This, on purification, is likewise a hygroscopic crystalline solid melting at 26°C. It is also optically active ($[\alpha]_D^{25} - 4°$). Lactic acid prepared from ethanal cyanohydrin is a hygroscopic low-melting solid (m.p. 18°C). It is not optically active. If exactly equal proportions of the lactic acid derived from muscle and the lactic acid derived by fermentation of milk are mixed together the solution is optically inactive and the solid obtained from the solution melts at 18°C. To distinguish between the two naturally occurring acids, that derived from muscle which rotates the plane of polarized light towards the right is called (+)-lactic acid (or *dextro*-lactic acid) while that obtained by fermentation of milk sugar is called (−)-lactic acid (or *laevo*-lactic acid). The lactic acid prepared from ethanal cyanohydrin or by mixing equal amounts of (+)-lactic acid and (−)-lactic acid is called inactive or *racemic* (±) lactic acid.

Conclusions

A compound which has molecules such that the mirror image of one molecule is not superimposable on that molecule is said to exhibit optical isomerism. The analogy with the pair of gloves should always be remembered. Any ordinary chemical synthesis of an asymmetric compound will produce the two mirror images in equal amounts and the resulting compound will not exhibit optical activity. If, however, it is possible to obtain a sample of that compound containing only one form (e.g. left-hand gloves only) then solutions of this sample will exhibit optical activity, i.e. they will rotate the plane of plane polarized light. Molecular asymmetry is particularly common in the chemistry of carbon compounds because a carbon atom with four different groups attached to it is asymmetric. However, molecular asymmetry is, as we have clearly indicated, by no means necessarily associated with the tetrahedral carbon atom.

Methods for separating enantiomers and what happens when there is more than one asymmetric centre in a molecule must await a more detailed discussion of polarized light and the experimental techniques involved.

Problems

1. Assuming you could separate the enantiomers, which of the following compounds would show optical activity?

(a) CHFClSO₃H

(b) CH₃CH₂CH(NH₂)CO₂H

(c)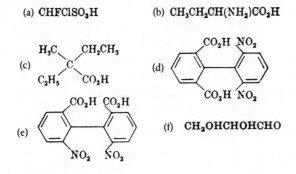

(d)

(e)

(f) CH₂OHCHOHCHO

2. Emil Fischer completed the following reaction sequence:

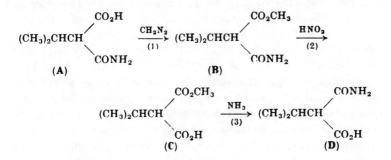

If compound A had a specific rotation of 49°40′ what was the specific rotation of compound D? (Reactions 1 and 2 have not been discussed in this book; reaction 3 is the normal aminolysis of an ester discussed in Chapter 9.)

CHAPTER 17 ——————————————

Naturally Occurring Organic Compounds

At the beginning of Chapter 1, we said that life is a complex series of chemical processes and these invariably involve the compounds of carbon. Many simple carbon compounds are involved in living processes, carbon dioxide occupying a key position in the carbon cycle. On the other hand, many extremely complex molecules are also involved; e.g. vitamin B_{12} has a molecular formula $C_{63}H_{90}O_{14}N_{14}PCo$, and all living organisms contain highly complex polymeric substances called proteins. Complete discussion of even the simplest type of naturally occurring compound is outside the scope of this book. This chapter will simply attempt to outline a few of the simpler classes of naturally occurring substances, for in spite of the complexity of many of them, they can often be regarded as belonging to a class which can be defined by some chemical feature. For example, the first class we consider are fats and oils which are esters of glycerol (propane-1,2,3-triol):

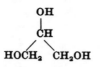

Fats and Oils

Just as an internal combustion engine depends on the energy released when hydrocarbons react with oxygen to yield carbon dioxide and water, so an animal obtains energy by the oxidation of its food. Although a great deal of food may be degraded directly, some is converted into compounds which can be stored in the body as a kind of reserve. One of the body's energy reserves is fat. Fat is a mixture of glyceryl esters, the formulae of which can be written in the following way:

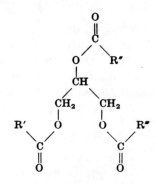

where R′, R″ and R‴ represent long straight hydrocarbon chains. These esters, occurring in animal and plant fats and oils, can be hydrolysed with hot aqueous sodium hydroxide to glycerol and the sodium salts of the long-chain acids:

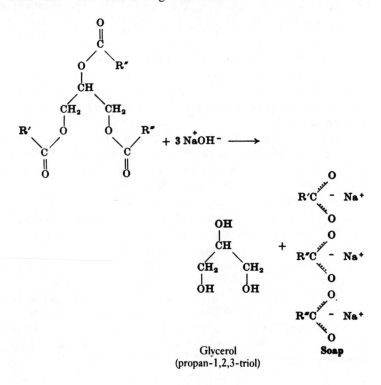

Glycerol
(propan-1,2,3-triol)

Soap

This process is used commercially to produce soap and is known as *saponification*.

Among the acids commonly obtained by the hydrolysis of fats are:

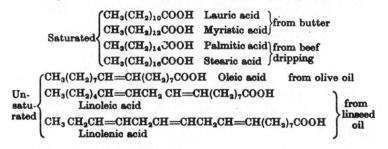

In general, plants and fish produce oils in which the glycerol esters contain unsaturated straight chains, while animal fat contains mainly saturated esters. The more unsaturated a glycerol ester is, the lower is its melting point, and in countries where vegetable oils are not widely used for cooking, these oils are hardened on a commercial scale by partial hydrogenation using hydrogen and a finely divided nickel catalyst (cf. Chapter 7):

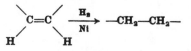

The hardened oil is used for making margarine and synthetic cooking fat.

Obesity, the deposition of excess fat in the body due primarily to overeating, has been associated with heart disease. There is a move away from the use of saturated fats—lard and dripping—in foods to the use of unsaturated fats—vegetable oils—on the grounds that there is a connection between the excessive consumption of animal, saturated fats and heart disease. However, a more moderate style of eating is probably more likely to lead to a decrease in the incidence of heart attacks.

Soap is made by the hydrolysis of both animal fat and hardened vegetable oil. The soap is separated from the aqueous glycerol solution by salting out with sodium chloride and filtering the precipitated sodium salts; these are pressed into cakes of soap.

The cleaning action of soap is a complex one involving surface forces. In simple terms it can be ascribed to the fact that the fatty acid anion has a long hydrocarbon tail attached to an ionic group. Dirt adheres to the skin and other surfaces mainly by films of oil.

The hydrocarbon chains are adsorbed by such oil films leaving the ionic carboxylate groups in the aqueous phase:

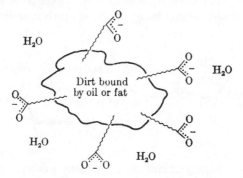

The presence of the carboxylate ion at the head of this tail renders the oil film partly soluble in water, enabling the oil and the dirt sticking to it to be dispersed to give a colloidal suspension, which may be rinsed away.

The waterproof nature of plant leaves is due to a surface layer of wax. This wax contains long-chain hydrocarbons (about C_{30}) as well as esters of long-chain monohydric alcohols (ROH) with long-chain acids ($R'CO_2H$).

Carbohydrates

A quite different class of organic molecules is used by plants and also by animals as energy reserves and by plants as their main structural building material. These compounds, many of which have the empirical formula $(CH_2O)_n$, are called *carbohydrates*. For example, plant seeds and roots contain reserves of energy in the form of starch which is a polymer having a complex structure which can be represented in the following way:

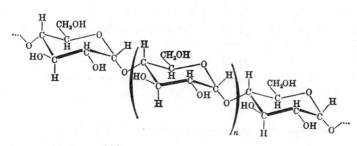

where *n* is about 300.

Sucrose (cane sugar), another energy reserve of plants, can be extracted from sugar cane and from sugar beet. It is a *disaccharide*, having two multihydroxy C_6 units joined together. Starch is called a *polysaccharide*, having many of the six-carbon units linked in a chain.

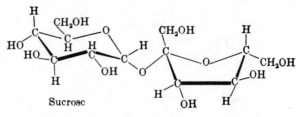

Sucrose

Cellulose is a polysaccharide formed by plants as a structural building material and is similar to starch in chemical make-up, differing principally in the stereochemistry of the link between the C_6 units and in the length of the chain. Both starch and cellulose can be hydrolysed and broken down into identical C_6 units, called glucose, a monosaccharide which also forms one half of the sucrose molecule and which occurs free in many living cells.

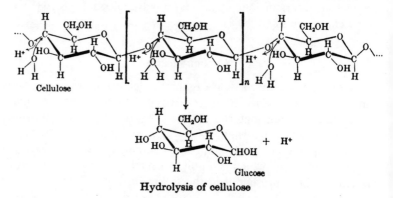

Hydrolysis of cellulose

This hydrolysis can be achieved with dilute mineral acids such as sulphuric acid. (This is why laboratory coats—made of cellulose fibres—are so easily rotted by acid splashes.) Hydrolysis can also be achieved by enzymes. The enzymic hydrolysis of poly-saccharides to monosaccharides such as glucose, followed by enzymic breakdown of glucose to ethanol and carbon dioxide (fermentation),

$$C_6H_{12}O_6(aq) \rightarrow 2\ C_2H_5OH(aq) + CO_2 + heat$$

is of widespread importance for the production of beer and other dilute solutions of ethanol which are used as beverages. It is also employed for the preparation of pure ethanol for the chemical industry.

To prepare industrial ethanol, starch, and sometimes cellulose, are hydrolysed by acids rather than enzymes. The formation of ethanol from glucose by fermentation is brought about by an enzyme in yeast. This step cannot be achieved by synthetic chemical reagents. The resulting dilute (10 per cent) solution of ethanol is concentrated by fractional distillation to give 'rectified spirit' which is an azeotrope of water and ethanol (95.6 per cent ethanol by weight). The remaining water is usually removed by addition of benzene and refractionating the mixture. The ternary mixture containing 18.5 per cent ethanol, 74.1 per cent benzene and 7.4 per cent water by weight distils at 65°C and permits complete dehydration if just sufficient benzene is added.

A dilute solution of ethanol, if exposed to air, is bacterially oxidized to a dilute ethanoic acid solution. This is the reason why wine and beer go sour if left uncorked, and it is the basis for the production of vinegar, the chief component of which is ethanoic acid.

If we look carefully at the structure of a typical monosaccharide such as glucose,

we see it as a hemi-acetal and it is thus sensitive to either acids or bases, and even in aqueous solution of pH 7 the following equilibrium is quickly established:

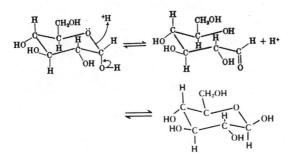

We would thus expect glucose to exhibit the properties of an aldehyde as well as those of a polyhydroxy compound, which it does. Most carbohydrates are asymmetric molecules and usually only one enantiomer is found in nature.

Proteins

Unlike plants, the structural material of animal tissues is proteins, which are polyamides of high molecular weight. These polyamides can be hydrolysed by acids, alkalis and enyzmes to give α-amino acids. Proteins have many functions besides their structural use and are important constituents of all living cells.

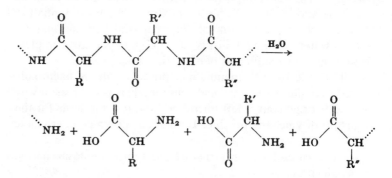

Simple proteins such as egg albumin and keratin from hair yield just α-amino acids, $RCHNH_2CO_2H$, when hydrolysed. About twenty such amino acids form the structural units of all the animal proteins that have been investigated. The molecular weights of proteins vary from 1.2×10^4 for insulin to 3×10^8 for influenza virus. Notice that the α-amino acids contain an asymmetric carbon atom and that only one enantiomer of any one amino acid occurs in a protein, and in general all naturally occurring amino acids have the same relative distribution in space of the four groups R, NH_2, H and CO_2H.

Silk is a fibrous protein produced by the silk worm. After several years of research in the 1930s, Carrothers of du Pont matched the skill of the silk worm and made a polyamide in the laboratory which had the properties of silk, called nylon:

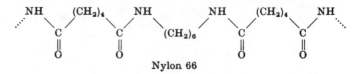

Nylon 66

Nylon 66 is prepared by the reaction of adipic acid (hexandioic acid) $HO_2C(CH_2)_4CO_2H$, with 1,6-diaminohexane, $H_2N-(CH_2)_6NH_2$.

Terpenes

Many plants and trees, particularly conifers, produce a class of organic compounds called terpenes. These compounds, often fragrant oils, are contained in the essential oils obtained by solvent extraction of plant blooms or steam distillation of leaves or resinified sap.

Rose oil contains citronellol:

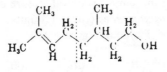

Citronellol

lemon oil contains limonene:

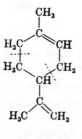

Limonene

and the principal component of oil of turpentine, which was formerly used as a paint thinner, is α-pinene:

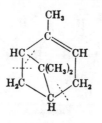

α-Pinene

All these compounds contain ten carbon atoms and exhibit a typical structure pattern which is based on two 5-carbon iso-pentane skeletons joined together:

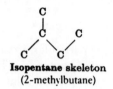

Isopentane skeleton
(2-methylbutane)

and are called *monoterpenes* (two C_5 units). The dashed lines on the formulae illustrate this.

Higher-boiling fractions of plant extracts contain C_{15} compounds exhibiting the same structural feature of the monoterpenes and are called *sesquiterpenes* (three C_5 units). Farnesol is present in many perfume oils including rose oil:

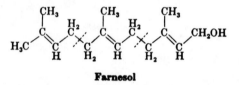

Farnesol

C_{20} compounds, called *diterpenes* (four C_5 units), also exhibit this structure pattern; vitamin A, a primary alcohol essential to the process of vision, may be isolated in high yield from fish oils:

C_{30} compounds, *triterpenes* (six C_5 units) and C_{40} compounds, *tetraterpenes* (eight C_5 units), are known. The latter, called *carotenoids*, constitute the yellow and red fat-soluble pigments of plants such as carrots and tomatoes. The structure of these compounds can be represented as dimers of vitamin A, and the body can convert some carotenoids into vitamin A; hence the belief that a diet of carrots improves vision.

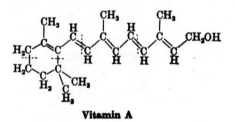

Vitamin A

Natural rubber is a *polyterpene* (about 10^4 C_5 units):

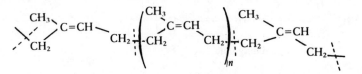

This polymeric molecule shares with many terpenes the property of giving isoprene when destructively distilled:

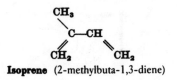

Isoprene (2-methylbuta-1,3-diene)

The structural similarity of all these different compounds containing the repeating isoprene unit joined head to tail was appreciated by organic chemists at the turn of the century who recognized the isoprene unit as a common feature present in most of the complex compounds in this group of natural products.

It has been shown that in living organisms isopentenyl pyrophosphate:

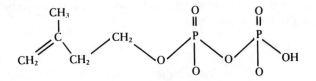

is the biologically active building unit derived from ethanoate ions which living organisms transform into terpenes and into steroids, a class of compound of great physiological significance which include sex hormones and compounds such as cortisone and heart poisons and also cholesterol, which has been associated with heart disease. Isopentenyl pyrophosphate itself is formed in living organisms from mevalonic acid:

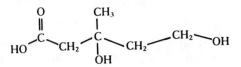

which in turn is derived from ethanoate ions.

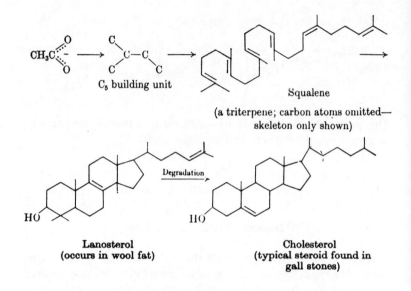

Lanosterol
(occurs in wool fat)

Cholesterol
(typical steroid found in
gall stones)

Alkaloids

A wide variety of basic substances containing nitrogen occur in plants. These compounds are loosely grouped together under the heading *alkaloids*. For example, opium (obtained from poppies) contains over twenty-four alkaloids, one of the most important medicinally being morphine:

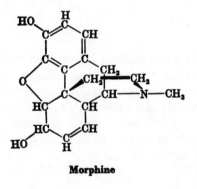

Morphine

The difficulty of attempting to draw in two dimensions a molecule which is not flat is even more obvious here than with the carbohydrates.

Nucleic Acids

Nucleic acids are an even more complex group of molecules that play a key part in the process of life. Nucleic acids are polymers that can have a molecular weight up to 4×10^9 with a repeat unit which is known as a nucleotide. Nucleotides are themselves made up of a phosphate unit, a sugar and a heterocyclic base. In ribonucleic acid (RNA) the sugar is ribose, a 5-carbon sugar; in deoxyribonucleic acid (DNA) the sugar is deoxyribose, a related 5-carbon sugar.

Deoxyribonucleic acid occurs in the nuclei of all cells and this molecule is responsible for storing genetic information for the replication of the cell by means of the sequence of bases attached to the sugar units:

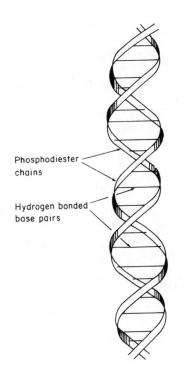

Figure 17.1 Model of the DNA double helix.

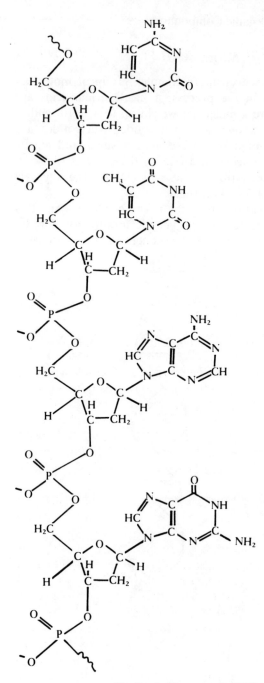

Base (cytidine)

Sugar (deoxyribose)

Phosphate

Base (thymidine)

Sugar (deoxyribose)

Phosphate

Base (adenine)

Sugar (deoxyribose)

Phosphate

Base (guanosine)

Sugar (deoxyribose)

Phosphate

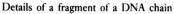

Details of a fragment of a DNA chain

The deoxyribonucleic acid exists in the cell as two polymer chains joined together, the 'double helix' (Figure 17.1), by hydrogen bonds formed between specific base pairs, the pairing being dictated by molecular geometry. For example, adenine pairs with thymine, guanine with cytosine:

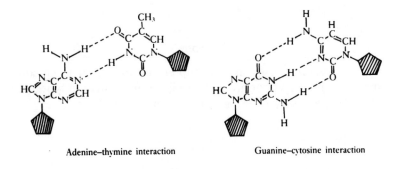

Adenine–thymine interaction   Guanine–cytosine interaction

When the cell divides (mitosis), the two hydrogen bonded chains separate and each separate chain acts as a template for the construction of another chain similar to the one with which it was paired before cell division, so that one double helix is converted into two, one for the old cell and one for the newly formed cell.

Ribonucleic acid has a similar structure to that of deoxyribonucleic acid, except that the sugar is ribose, some of the bases are different and ribonucleic acid exists as single strands. Ribonucleic acid acts as a template for the synthesis of proteins in the cell.

Thus the process of life is dependent on the interactions of these large molecules, and these interactions in turn are dependent on the simple hydrogen bond.

Organic chemistry has been very much concerned with the determination of the structure and subsequent synthesis of natural products, only a few of which have been mentioned in this section. In recent years attention has been focused on how the plant or animal makes these complicated compounds.

Organic compounds produced by living organisms are usually too complex to serve as raw material for the production of simple organic compounds to form the basis of organic chemical synthesis. The one notable exception is the formation of ethanol from such carbohydrates as cellulose or starch.

For the production of organic compounds on a scale vast enough to satisfy the requirements of industry, we must turn to substances which were produced by living organisms many millions of years ago, and which, during the course of time, have been modified by heat and pressure. In the next chapter we will describe the source of the many compounds that are used by the chemical industry.

Note

The structures given in this chapter are not meant to be committed to memory, but are given to illustrate structural types and chemical reactivity of the groups of natural products.

Problems

1. Show, by means of equations, the chemistry of the following processes:
(a)   The production of soap and glycerol from a liquid vegetable oil
  (b) The production of ethanol from starch

2. Ozonization of natural rubber and hydrolysis of the ozonide leads to the formation of laevulinic aldehyde (pentan-4-onal, $CH_3CCH_2CH_2CH=O$) in about 95 per cent yield. Natural rubber
$$\overset{\parallel}{\underset{O}{}}$$
adds 1 molecule of H_2 for each five carbon atoms of the molecule, and yields isoprene when destructively distilled. Show how these facts can be used to derive a structure for natural rubber.

3. Show how the following compounds may be classified by dividing into isopentane units:

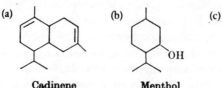

(a)   Cadinene
(from cade oil)

(b)   Menthol
(from mint oil)

(c)   Eudesmol
(from eucalyptus oil)

CHAPTER 18 —————————————————

Industrial Sources of Organic Compounds

In Chapters 3 to 15 we discussed the reactions of various types of organic molecule without considering from where such compounds might be derived. In the last chapter we considered a few examples of the vast range of organic compounds present in living organisms. However, it was indicated that industry does not obtain the many simple organic compounds needed for synthetic purposes from complicated natural products. The main source is petroleum, believed to be the remains of prehistoric living material.

At the end of the nineteenth century, the production of organic chemicals from the coal industries of European countries was well established, but by the second decade of the twentieth century the petrochemical industry was developing in the United States of America. This chemical industry was derived from the products of the oil industry, serving the needs of the internal combustion engine. The oil-based industry spread throughout the world as oil reserves were discovered outside the United States of America. The relative ease with which oil can be extracted from the earth and the ease with which it can be handled compared to coal has led to the displacement of coal as a fuel and energy source, and consequently as a source of organic chemicals. In 1950, 40 per cent of the free world's production of organic chemicals was based on oil; in 1985 it was estimated that more than 99 per cent of organic chemical production is oil based.

Petroleum

In this century the emphasis has shifted from the use of coal as source of energy for industry to the other major fossil fuel,

219

petroleum, which is more easily handled. Petroleum is believed to be formed by the decomposition of organic material, possibly marine in origin, and the products of this decomposition have accumulated in porous strata subsequently capped by impervious layers of rock. High pressures and temperatures have probably not been a part of the process forming petroleum, as is the case in coal formation.

The formation of petroleum is probably analogous to the process which can be observed in any stagnant pond where rotting vegetation gives off methane, CH_4 (marsh gas). In fact the treatment of sewage results in the formation of sufficient methane to provide power to fire the boilers of some sewage disposal works.

Refineries are familiar parts of the present-day industrial landscape. What does petroleum consist of and what are the chemical reactions that go on in these refineries?

Petroleum contains a mixture of hydrocarbons, mainly alkanes, ranging from CH_4 (the major constituent of natural gas) to compounds containing over 100 carbon atoms. Some petroleums contain a small proportion of cycloalkanes; others, found in central Europe, contain some 5 per cent or so of aromatic hydrocarbons. The first operation that petroleum is subjected to is that of fractional distillation. In the fractionating column the petroleum is separated roughly according to molecular weight into fractions of increasing boiling point (Table 18.1).

Let us now look at some of the reactions and the utility of these fractions.

Table 18.1   Fractions obtained from the distillation of crude oil.

Fraction	Boiling range (°C)	Range of number of C atoms in the constituent hydrocarbons
Gas	<20	C_4
Gasoline (naphtha)	20–200	C_4–C_{12}
Kerosene	ca. 175–270	C_9–C_{16}
Gas oil (middle distillates)	ca. 200–400	C_{15}–C_{25}
Lubricating oil	ca. 300 (in vacuum)	C_{20}–C_{30}
Fuel oil	—	> C_{30}
Residue	—	—

Natural Gas

Natural gas consists of mainly CH_4 and C_2H_6 with traces of C_3H_8 and C_4H_{10}. It is burnt as a fuel, and at present has replaced town or coal gas that used to be obtained by the carbonization of coal:

$$CH_4(g) + 2\,O_2\,(g) \longrightarrow$$
$$CO_2(g) + 2\,H_2O\,(l) \quad \varDelta H = -212{\cdot}8 \text{ kcal mol}^{-1}$$

The combustion can be controlled to produce:
(a) Methanol

$$CH_4 + \tfrac{1}{2}\,O_2 \longrightarrow CH_3OH$$

which is also produced by synthesis:

$$CO + 2\,H_2 \xrightarrow[\text{High pressure}]{\text{Oxide catalyst}} CH_3OH$$

(b) Formaldehyde

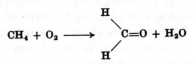

$$CH_4 + O_2 \longrightarrow \quad \overset{H}{\underset{H}{\diagdown\diagup}}C{=}O + H_2O$$

(c) Carbon black, finely divided carbon particles used in tyres to improve the mechanical properties of the rubber

$$CH_4 + O_2 \longrightarrow C + 2\,H_2O$$

(d) Or it is 'cracked' to serve as a source of hydrogen:

$$CH_4 \xrightarrow[\substack{\text{decomposition} \\ \text{at } 1200°C}]{\text{Thermal}} C + 2\,H_2$$

These are all very complex reactions and the equations only summarize the overall process. It is thought that the reactions are free radical in nature.

Petrol

The main use for petrol is in spark-ignition engines. It is found that straight-chain hydrocarbons such as heptane, $CH_3(CH_2)_5CH_3$

('octane value' = 0), give rough running with 'pinking' or 'knocking' under load in a spark-ignition engine. This characteristic pinking can be ascribed to the explosive detonation, rather than the smooth combusion, of the fuel–air mixture. Branched-chain hydrocarbons such as the so-called 'isooctane',

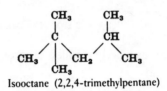

Isooctane (2,2,4-trimethylpentane)

(octane rating = 100), give very smooth running even in engines with a high compression ratio. The octane number or rating of a petrol is defined as the proportion of isooctane which has to be added to heptane to match the running characteristics of the petrol under test.

Tetraethyl lead, $(C_2H_5)_4Pb$, has been extensively used since the 1920s as an additive to improve the octane rating of petrol. The tetraethyl lead suppresses knocking by forming a fog of lead(II) oxide (PbO) in the combustion chamber, which acts as an inhibitor to the uncontrolled oxidative chain reaction which results in premature detonation of the air–fuel mixture. This fog of PbO must not be allowed to collect on the valves and spark plug of the engine, so that 1,2-dibromoethane, $BrCH_2CH_2Br$, is also added to the petrol, to form volatile $PbBr_2$, which passes out through the exhaust into the atmosphere.

Anxiety has recently been expressed as to the effect of this lead emission into the atmosphere on the population, especially children, living near heavily used roads, and the presence of traces of lead in the body has been associated with irreversible brain damage. A number of countries have now banned the use of tetraethyl lead in petrol, thus increasing the demand for refineries to produce branched-chain hydrocarbons.

Unlike spark-ignition engines, the reverse is true for diesel (compression-ignition) engines. Here, straight-chain hydrocarbons give smoother running than branched-chain hydrocarbons and different standard fuels are used to rate diesel fuels. Not enough branched-chain hydrocarbons were available from the primary fractional distillation of crude petroleum to make high-octane petrol, and refineries now operate several reactions to turn unwanted straight-chain hydrocarbons into branched-chain hydrocarbons, and also to convert surplus high-boiling fractions (fuel oils) into more volatile hydrocarbons.

a. Isomerization

At room temperature, branched-chain hydrocarbons are thermodynamically more stable than straight-chain hydrocarbons. $AlCl_3$ is used to catalyse this isomerization which takes place by way of a hydride shift, and involves a carbocation intermediate and the migration of a methyl group with an electron pair:

$$CH_3CH_2CH_2CH_2CHCH_2CH_3 \rightleftharpoons CH_3CH_2CH_2CH_2\overset{+}{C}HCH_2 \frown CH_3 + [HAlCl_3]^- \rightleftharpoons$$

Heptane AlCl₃

H
AlCl₃
Heptane

$$\underset{+}{CH_3CH_2CH_2CH_2CH}\overset{CH_3}{\overset{|}{—CH_2}} \xrightarrow{[HAlCl_3]^-} CH_3CH_2CH_2CH_2CHCH_3 + AlCl_3$$

CH₃

2-Methylhexane

The reaction is more complex than has been indicated, as traces of HCl and alkenes are essential for the reaction to start.

b. Alkylation

Branched-chain hydrocarbons are prepared from alkenes, e.g. isooctane from 2-methylpropene:

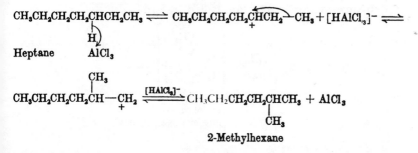

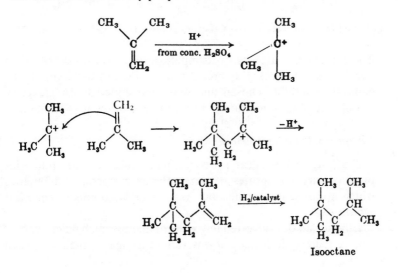

Isooctane

The reaction involves the addition of one C_4 unit to another, catalysed by a strong acid (see Chapter 7). The resulting branched-chain alkene is hydrogenated. Alkenes have high octane ratings but are too reactive, and may polymerize, so clogging pipes and carburettors.

c. Reforming

Aromatic hydrocarbons have high octane values and can be used in petrols. They are obtained from alkanes containing from six to seven carbon atoms available in the refinery by cyclo-dehydrogenation, a reaction called 'reforming':

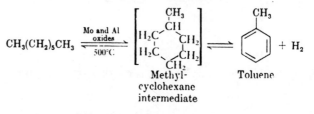

Methyl-
cyclohexane
intermediate

Toluene

d. Thermal Cracking

This process is used to turn surplus high-boiling fractions into low-boiling petrol fuels. It is a free-radical thermal decomposition, giving many products, and involves the breaking of carbon–carbon bonds:

$$CH_3(CH_2)_3CH_2\frown CH_2(CH_2)_3CH_3 \xrightarrow[\text{450–500°C}]{\text{Al, Si oxides}} 2\ CH_3(CH_2)_3\dot{C}H_2$$

$$2\ CH_3(CH_2)_2CH_2\dot{C}H_2 \longrightarrow CH_3(CH_2)_2CH{=}CH_2 + CH_3(CH_2)_2CH_2CH_3$$

The equation is a gross oversimplification as the carbon–carbon bond does not necessarily break in the middle of the chain. If the catalyst is suitably chosen and hydrogen added to the reaction vessel, the alkene mixture is reduced as soon as it is formed.

Chemicals from Petroleum

Cracking reactions can produce small molecules, and refineries produce large quantities of pure ethene, propene and butenes which are used not as fuels but as the starting points of chemical syntheses.

Ethene can be converted into chloroethane, which when reacted with a lead/sodium alloy forms tetraethyl lead which was used as an anti-knock additive to petrol, but is now regarded as an environmental hazard:

$$CH_2{=}CH_2 \xrightarrow{\text{HCl}} CH_3CH_2Cl \xrightarrow{\text{Pb/Na}} (C_2H_5)_4Pb$$

Ethene is converted by H_2SO_4 followed by hydrolysis into ethanol (Chapter 7), used by industry as a solvent and as a starting material for other solvents such as esters:

$$CH_2{=}CH_2 \xrightarrow{\text{H}_2\text{SO}_4} CH_3CH_2OSO_2OH \xrightarrow{\text{H}_2\text{O}} CH_3CH_2OH$$

Ethene itself is the monomer of poly(ethene):

$$n\ CH_2{=}CH_2 \xrightarrow[\text{R.X}]{\text{Initiator}} R{-}(CH_2CH_2)_n{-}X$$

a plastic used on a vast scale for domestic articles such as washing-up basins and squeezable bottles (Chapter 7). Poly(propene), made by polymerization of propene, has better mechanical strength and a higher melting point than poly(ethene). It is used in the manufacture of sacks, ropes and carpet backing.

Several other organic molecules of everyday importance, such as glycol (ethane-1,2-diol), are made from ethene by simple reactions (Chapter 7).

Ethyne, an important starting material for the syntheses of ethanoic acid, vinyl chloride and acrylonitrile, is obtained on a large scale by the thermal decomposition of ethane or, more usually, a mixture of low-boiling alkanes. At high temperature the following equilibrium is established:

$$CH_3{-}CH_3 \underset{1500^\circ\text{C}}{\overset{1000-}{\rightleftharpoons}} CH{\equiv}CH + 2\ H_2$$

The alkanes can be converted either by partial combustion in a restricted supply of air or by passing the alkanes through an electric arc. The conversion only occurs to the extent of 5 per cent but the ethyne can be removed from the issuing gases and unchanged alkanes recycled. This process is competitive with the calcium carbide method for producing ethyne:

$$\text{Coke} + \text{CaO} \xrightarrow[\text{furnace}]{\text{Electric}} \underset{\substack{\text{Calcium}\\\text{carbide}}}{CaC_2} \xrightarrow{2\text{H}_2\text{O}} CH{\equiv}CH(g) + Ca(OH)_2$$

Both vinyl chloride and acrylonitrile are monomers for plastics. Polyvinyl chloride is used as a substitute for leather, in long-playing records and for plastic baby pants. Ethyne is no longer used as a starting material for the manufacture of vinyl chloride as the addition of hydrogen chloride to ethyne needs to be catalysed by mercury salts. An instance of the careless disposal of waste from a vinyl chloride plant has resulted in disastrous environmental pollution due to the inclusion of poisonous mercury salts into a food chain eventually leading to man.

Vinyl chloride is now made by the simultaneous oxidation and chlorination of the cheaper ethene:

$$CH_2 = CH_2 + \tfrac{1}{2}O_2 + HCl \xrightarrow{\text{Catalyst}} CH_2{=}CHCl + H_2O$$

Polyacrylonitrile is used in the production of artificial fibres such as 'Acrilan'.

Arene Derivatives from Petroleum

Petroleum is now used as a source of aromatic compounds, e.g. the preparation of phenol on an industrial scale from benzene and propene. The reactions involving alkylation, followed by oxidation. Propanone, an important solvent, is a useful by-product of this reaction.

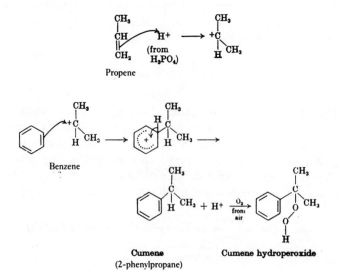

Benzene

Cumene
(2-phenylpropane)

Cumene hydroperoxide

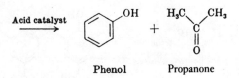

Phenol   Propanone

The artificial polyester fibre 'Terylene' is made from *p*-xylene (1,4-dimethylbenzene) and ethan-1,2-diol:

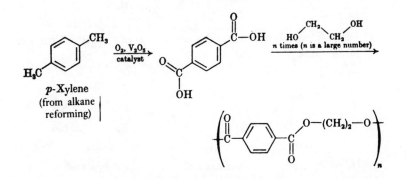

p-Xylene
(from alkane
reforming)

The diagram represents the overall course of the reaction. In practice it is difficult to obtain terephthalic acid sufficiently pure for the polymerization stage, and the dimethyl ester of terephthalic acid is used. This ester can be more easily purified, and it reacts with ethan-1,2-diol to form Terylene, with methanol a by-product.

The azo dye 'Para-red' is made by treating diazotized *p*-nitro-aniline with the sodium salt of 2-naphthol:

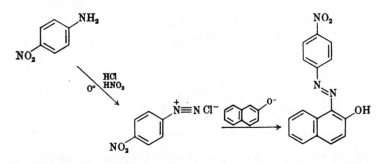

The cross-linked polymer 'Bakelite', which was used in the electrical industry and is still used for laminating wood, is made from phenol and formaldehyde:

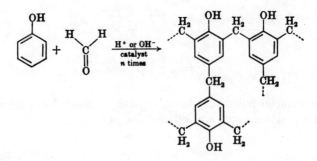

Phenol or cyclohexane may be used as a raw material for the polyamide fibre 'Nylon 66':

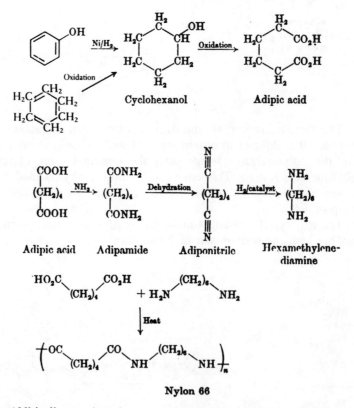

Nylon 66

The '66' indicates that there are six carbon atoms in both the acid and amine component. Other useful nylons are manufactured which contain a different number of carbon atoms in both fragments.

Detergents from Petroleum

Synthetic detergents are usually the sodium salts of long-chain sulphonic acids. These detergents are often more successful cleansing agents in hard-water areas than soap, as their Ca^{2+} and Mg^{2+} salts are soluble in water in contrast to the Ca^{2+} and Mg^{2+} salts of the fatty acids in soap, the latter being insoluble in water.

A common type of detergent, used in household washing powders and liquids, is prepared by alkylating benzene with an alkyl halide or terminal alkene (R about C_{12}), sulphonating the product and treating this with sodium carbonate:

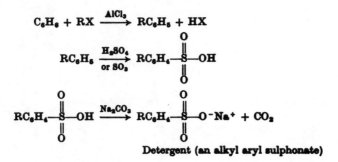

Detergent (an alkyl aryl sulphonate)

If R, the alkyl chain, is branched, these detergents are resistant to degradation by microbes and have caused problems in sewage works, where discharge of undegraded detergents into outflows has led to unacceptable levels of foaming in streams and rivers. As a consequence most detergents now contain straight, unbranched alkyl chains, which are more expensive to produce but which are necessitated by popular concern about protecting our environment.

This chapter has presented an incomplete survey intended to show how petroleum can be used as a source of small, reactive molecules which are in turn used for starting materials for industrial syntheses. Some of the reactions described here appear to be different from those discussed previously. This is because in industrial plant very high temperatures and pressures can be used and reactions which are improbable in the ordinary laboratory can become feasible. However, all these reactions do in fact involve the principles we have discussed earlier, and many of the reactions are exactly the same as those performed in the laboratory but are merely scaled up to produce tons of product instead of grams.

Problems

1. Indicating where possible the type of reaction involved, show the steps involved in forming the following from petroleum:

(a) Tetraethyl lead   (d) Toluene

(b) Polyethylene   (e) Polyvinyl chloride

(c) Ethanol alcohol   (f) Ethanoic acid

2. Show how petroleum serves as a source of starting compounds for the production of:

(a) Nylon 66   (c) Bakelite

(b) Terylene   (d) An azo dye

CHAPTER 19 ———————————

Synthesis

Chapters 3 to 15 have been concerned with the chemical reactions of different types of bonds and groupings attached to a carbon chain. Chapters 17 and 18 gave an account of the natural sources from which carbon compounds are obtained. It is the purpose of this chapter to draw together all the reactions we considered earlier in order to see how we can convert one carbon compound into another, bearing in mind always the kind of compounds we have to start with from natural sources. Many textbooks have lists of methods of preparation for each class of organic compound. It should already be clear, however, that with the possible exceptions of methanol and ethyne, we would never, in a strict sense, prepare any carbon compound. If we want a molecule which is not readily available from some natural source such as petroleum, we take a petroleum product as closely related as possible and convert it by a series of chemical reactions into the compound we desire. Thus, this chapter is really not concerned with the preparation but with the interconversion of different types of groupings. Very few new reactions will be introduced and those which are will not involve any new principles. We will work through the different types of bonds and groupings in the same order as they have been discussed, considering in each case how a particular grouping can be introduced into an organic molecule. When a reaction has been described in a previous chapter a reference will be given to the appropriate chapter and, in such cases, the electron transfers will not be detailed, since these can be found by looking up the reference.

The Carbon–Hydrogen Bond

It should be clear from Chapter 18 that, in general, there is no need to attempt to synthesize hydrocarbons. These are readily available from petroleum, although to obtain one particular hydrocarbon free from any of the others may be a difficult business. There may be occasions when we wish to remove a particular grouping from a molecule and replace it by a hydrogen atom and we shall briefly consider ways in which this can be done.

1. A carbon–halogen bond can be replaced by a carbon–hydrogen bond via the Grignard reaction (see Chapter 13):

$$RCH_2Br \xrightarrow[\text{(C}_2\text{H}_5\text{)}_2\text{O}]{\text{Mg}} \underset{\substack{\text{Grignard} \\ \text{reagent}}}{RCH_2MgBr} \xrightarrow{\text{H}_2\text{O}} RCH_3$$

2. There are three reactions by which we can convert a carbonyl group into a methylene group. The first of these, described in Chapter 8, is the Wolff–Kishner reaction:

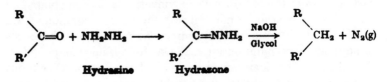

A closely related reaction described in Chapter 8 involves the reaction of the carbonyl compound with a thiol, followed by treatment of the thioacetal with hydrogen and nickel:

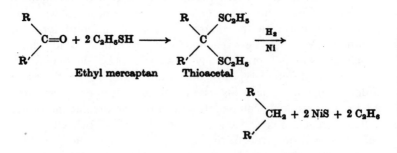

A third method of carrying out this transformation via the Clemensen reaction is described in Chapter 8 and involves the reaction of the carbonyl compound with hydrochloric acid and amalgamated zinc:

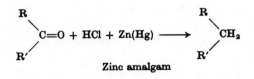

3. Alkenes and alkynes can of course be converted into saturated hydrocarbons by hydrogenation, which is described in Chapter 7 for alkenes and Chapter 12 for alkynes.

The Carbon–Halogen Bond

Alkyl halides are reagents of great value in synthesis and therefore methods of obtaining them from other compounds are very important.

1. The replacement of an alcoholic hydroxyl group by a halogen can be carried out in a number of ways. These are discussed in Chapter 5

$$RCH_2OH + HBr \longrightarrow RCH_2Br + H_2O$$

It is important to remember that the reaction between an alcohol and hydrogen chloride or hydrogen bromide requires the anhydrous hydrogen halide and not its aqueous solution. It is also important to remember that a halide anion will not replace a hydroxyl group from an alcohol but that the reaction involves the halide anion replacing the water molecule in a protonated alcohol or oxonium ion. This is discussed in Chapter 5. In place of the free acid we can use inorganic acid halides such as phosphorus halides or thionyl chloride:

$$3 ROH + PBr_3 \xrightarrow{\text{Pyridine}} 3 RBr + H_3PO_3$$

$$ROH + SOCl_2 \xrightarrow{\text{Pyridine}} RCl + HCl + SO_2$$
Thionyl
chloride

2. The carbon–halogen bond is readily formed by the addition of a hydrogen halide to an alkene (see Chapter 7):

$$RCH{=}CHR + HF \longrightarrow RCH_2CHFR$$

3. Carbon–halogen bonds are formed in a reaction, not previously discussed, in which the silver salt of a carboxylic acid and a halogen (especially Br_2 or I_2) are heated together in boiling carbon tetrachloride. This reaction probably involves free radicals.

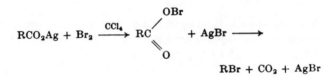

Notice that in this reaction we have lost one carbon atom, breaking a carbon–carbon bond, so that we can consider this reaction as a method of shortening the carbon chain.

4. The direct reaction of molecular halogen with a hydrocarbon is not an important method for making a carbon–halogen bond in the laboratory because such a reaction normally results in the formation of a number of isomers. If, for example, we cause propane to react with chlorine, initiating the reaction by light or by heat, we obtain a mixture of 1- and 2-chloropropane (see Chapter 3):

$$CH_3CH_2CH_3 \xrightarrow[\text{light}]{Cl_2} CH_3CHClCH_3 \text{ and } CH_3CH_2CH_2Cl$$

5. In many cases it is possible to change one carbon–halogen bond by another, which is described in Chapter 4:

$$RCl + KF \xrightarrow[\text{glycol}]{Dry} RF + KCl$$

$$RCl + NaI \xrightarrow{Acetone} RI + NaCl$$

With the exception of reaction 3, i.e. the reaction of the silver salt of a carboxylic acid with molecular halogen in carbon tetrachloride, none of the other reactions described above can be used for making an aromatic carbon–halogen bond. We have described two reactions by which this can be done (Chapter 15), namely:

6. Chloro- and bromobenzene can be prepared by the direct

reaction of benzene with the halogen and a suitable metal halide catalyst (Chapter 15):

$$C_6H_6 + Br_2 \xrightarrow{FeBr_3} C_6H_5Br + HBr$$

7. The amino group in aniline can be replaced by a halogen via the diazonium salt as described in Chapter 15:

$$C_6H_5NH_2 \xrightarrow[HCl]{HNO_2} C_6H_5\overset{+}{N}\equiv N\ Cl^- \xrightarrow{CuCl} C_6H_5Cl + N_2$$

The Hydroxyl Group

We have seen in Chapters 17 and 18 that methanol, ethanol and propan-2-ol are produced industrially on a large scale and so we must think of these compounds as starting points from which we can attempt to synthesize other molecules and not as compounds that we should ever want to prepare. The introduction of OH groups is, however, often of great importance in more complex molecules.

1. The replacement of a halogen by a hydroxyl group is described in Chapter 4. This is the classic example of a displacement reaction. For preparative purposes this reaction works best with primary alcohols where the danger of an accompanying elimination reaction (see Chapter 6) is less likely:

$$RBr + OH^- \longrightarrow ROH + Br^-$$

2. Ethanol is prepared industrially by the hydration of ethene (see Chapter 7) and this hydration is a reaction which can be of general synthetic importance:

$$RCH{=}CHR + H_2SO_4 \longrightarrow \underset{OSO_3H}{RCH_2CHR} \xrightarrow{H_2O} \underset{OH}{RCH_2CHR}$$

Closely related to this reaction is the reaction of alkenes with aqueous potassium permanganate or with osmium tetroxide to produce vicinal dihydroxy compounds, i.e. glycols:

$$RCH{=}CHR \xrightarrow[H_2O]{KMnO_4} \underset{OH\ \ OH}{RCH{-}CHR}$$

3. Alcohols can be prepared by the reduction of carbonyl compounds, aldehydes yielding primary alcohols and ketones yielding secondary alcohols (see Chapter 8):

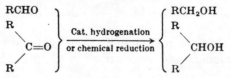

4. Esters can also be reduced to primary alcohols although, as discussed in Chapter 10, reduction of carboxylic acid derivatives is not normally a satisfactory reaction (see Chapter 10):

$$RCO_2C_2H_5 \xrightarrow[\text{or Na} + C_2H_5OH]{\text{LiAlH}_4} RCH_2OH$$

5. Two methods for the industrial preparation of phenol have been described (see Chapters 15 and 18). Neither of these processes is practical in the laboratory but aniline can be converted into phenol via the diazonium salt (see Chapter 15):

$$C_6H_5NH_2 \xrightarrow[\text{HCl}]{\text{HNO}_2} C_6H_5\overset{+}{N}\equiv N\ Cl^- \xrightarrow[\text{Heat}]{\text{H}_2\text{O}} C_6H_5OH$$

The Amino Group

It is not possible to prepare a primary amine by the reaction of ammonia and an alkyl halide because, as described in Chapter 4, the reaction goes further and the products of such a reaction are a mixture of primary, secondary and tertiary amines together with some of the quaternary salt. Similarly, it is not possible to prepare a primary aliphatic amine by the reaction of an alkyl halide with sodamide because the amide anion, NH_2^-, is a very powerful base and the result of such a reaction will be elimination rather than displacement (see Chapter 6).

1. In Chapter 9 we described how amides were amphoteric substances and how, acting as weak acids, they would form salts with the alkali metals. Benzene-1,2-dicarboxylic acid is called *phthalic acid* and although this compound will form a normal diamide called *phthalamide*, it more readily forms an imide in which the nitrogen atom is attached to two carbonyl groups:

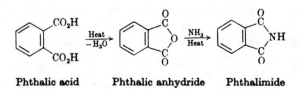

Phthalic acid   Phthalic anhydride   Phthalimide

Phthalimide is very much more acidic than ethanoamide and readily forms a potassium salt with potassium hydroxide or even potassium carbonate. The phthalimide anion is a suitable electron donor with which to react an alkyl halide. The product of this reaction, an N-alkylphthalimide, can then be hydrolysed to yield a primary amine and phthalic acid:

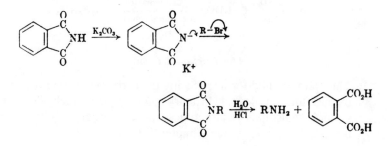

This reaction is known as the *Gabriel phthalimide synthesis* of amines.

2. Primary amines can also be prepared by the reduction of nitriles (see Chapter 12), but the reaction may be accompanied by undesirable side reactions giving secondary and even tertiary amines in addition to the desired primary amine. It is used industrially in the preparation of 1,6-diaminohexane (see Chapter 18):

$$RCN \xrightarrow[\text{or LiAlH}_4]{\text{Cat. H}_2} RCH_2NH_2$$

Likewise, amides can be hydrogenated and this enables us to prepare secondary or tertiary amines from secondary or tertiary amides:

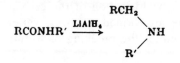

3. Amines can also be prepared by the hydrogenation of oximes or hydrazones formed by the reaction of aldehydes and ketones with hydroxylamine or hydrazine derivatives:

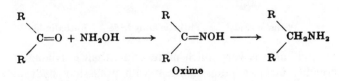

Oxime

4. Aniline cannot be prepared by any of the above reactions, but is readily formed by the hydrogenation of nitrobenzene (see Chapter 15):

$$C_6H_6 \xrightarrow[H_2SO_4]{HNO_3} C_6H_5NO_2 \xrightarrow[\text{or chemical reduction}]{\text{Cat. reduction}} C_6H_5NH_2$$

The Carbonyl Group

1. The main method for introducing the carbonyl group is by the oxidation of an alcohol. Primary alcohols yield aldehydes and secondary alcohols yield ketones whereas tertiary alcohols are not readily oxidized (see Chapter 11):

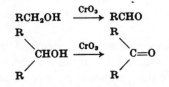

2. An acetylenic group can be converted into a methylene group and an adjacent carbonyl group; this is the way in which ethanal can be manufactured from ethyne (see Chapter 12):

$$RC{\equiv}CR \xrightarrow[H_2SO_4]{Hg^{2+}} RCH_2COR$$

These two reactions are the only simple conversion reactions which result in the formation of the carbonyl group. However, carbonyl groups are formed in a wide variety of other reactions which involve either the breaking or forming of carbon–carbon bonds.

3. An example of the formation of a carbonyl group as the result of a degradation of breaking of a carbon chain is the ozonolysis of an alkene, described in Chapter 7:

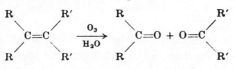

4. Ketones are formed in the reaction of Grignard reagents with esters, but the ketone so formed then reacts further with more Grignard reagent to yield the tertiary alcohol. The reaction of Grignard reagents with nitriles, however, can be used to prepare ketones (see Chapter 13):

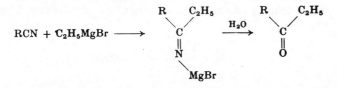

Grignard reagents react with acid chlorides to yield ketones but, as in the case of esters, the resulting ketone then reacts further with more Grignard reagent to yield the tertiary alcohol. A method of avoiding this difficulty is to use a less reactive organometallic derivative. Grignard reagents react with cadmium chloride to yield a dialkylcadmium. The dialkylcadmium then reacts with an acid chloride to yield the corresponding ketone:

$$2\ C_2H_5MgBr + CdCl_2 \longrightarrow (C_2H_5)_2Cd + MgBrCl$$
$$(C_2H_5)_2Cd + 2\ RCOCl \longrightarrow 2\ RCOC_2H_5 + CdCl_2$$

5. Phenyl ketones can be prepared by the reaction of an acid chloride and aluminium chloride in the Friedel–Crafts acylation reaction (see Chapter 15):

$$C_6H_6 + CH_3COCl + AlCl_3 \longrightarrow C_6H_5COCH_3$$
$$\text{Acetophenone}$$

The Carboxyl Group

Ethanoic acid is a commercial product produced in large quantities from the oxidation of ethanal derived from ethyne, or by the oxidation of ethanol from ethene, or by the direct oxidation of short-chain hydrocarbons from the petroleum refinery. Long-chain carboxylic acids are obtained by the hydrolysis of fats, as described in Chapter 17, and by the oxidation of long-chain hydrocarbons.

1. The main method of introducing a carboxyl group into a straight carbon chain is by oxidation of the appropriate primary alcohol to the aldehyde which is then oxidized to the carboxylic acid (see Chapter 8):

$$RCH_2OH \longrightarrow RCHO \longrightarrow RCO_2H$$

2. Another very important way of preparing carboxylic acids is the hydrolysis of a cyanide by either acid or base (see Chapter 12):

$$RCN \xrightarrow[\text{Acid or base}]{H_2O} RCO_2H$$

3. Of equal importance is the conversion of an alkyl halide into a carboxylic acid containing one more carbon atom via the Grignard reagent (see Chapter 13):

$$RBr \xrightarrow[(C_2H_5)_2O]{Mg} RMgBr \xrightarrow{CO_2} RCO_2H$$

It is a good opportunity, while we are discussing the carboxyl group, to consider how we can make esters and amides. The direct interaction of a carboxylic acid and an alcohol is discussed in Chapter 9, and this section should be carefully re-read. It will be clear from this discussion that the direct preparation of an ester from an alcohol and a carboxylic acid is a somewhat limited reaction. The alternative is to make the acid halide or the acid anhydride (see Chapter 10). This is the only way in which tertiary alcohols can be esterified and also the only way in which phenol can be esterified. The method of preparing esters of very limited application was described in Chapter 4, and involves the reaction of the sodium salt of the carboxylic acid with an alkyl halide. Primary amides can be prepared from esters and ammonia as described in Chapter 9. The 'acylation' of primary and secondary amines to secondary and tertiary amides requires the use of either an acid chloride or an acid anhydride.

The Nitrile Group

1. The most important route to an alkyl cyanide is to use an alkyl halide and treat this with potassium cyanide (Chapter 2):

$$RBr + KCN \xrightarrow{C_2H_5OH} RCN + KBr$$

α-Hydroxy cyanides are prepared by the reaction of an aldehyde or ketone with aqueous hydrogen cyanide (Chapter 8):

$$RCHO + HCN \longrightarrow RCHOHCN$$

2. Nitriles can also be prepared by the dehydration of amides in a reaction that we have not described previously. This simply involves treating the amide with a powerful dehydrating agent such as phosphorus pentoxide:

$$RCONH_2 \xrightarrow[P_2O_5]{-H_2O} RCN$$

The Carbon–Carbon Double Bond

Elimination reactions and the formation of alkenes form the subject of Chapter 6; a brief discussion of when displacement and when elimination are likely to occur is given on page 76 of that chapter. An important point is that the stronger the base, the more likely reaction with an alkyl halide and a sodamide results in elimination whereas the reaction between sodium ethanoate and an alkyl halide almost invariably results in displacement. The structure of the alkyl halide is also extremely important and the more branched the alkyl halide, the more likely is elimination. From a practical point of view, the most convenient way of making alkenes is from alcohols, either by direct dehydration as described on page 80 or (better) by way of the ester, particularly the xanthate esters as described on the same page.

$$RCH_2CH_2OH + CS_2 \xrightarrow{NaOH} RCH_2CH_2OCS_2^- + Na^+$$
Sodium xanthate ester

$$RCH_2CH_2OCS_2^- Na^+ \; CH_3{-}I \longrightarrow RCH_2CH_2OCS_2CH_3 + Na^+I^-$$

Methyl xanthate ester

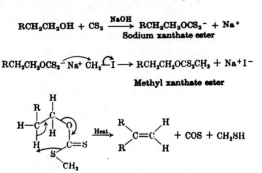

The Hofmann elimination described in Chapter 6 is also of considerable practical importance:

$$RCH_2CH_2NH_2 \xrightarrow{\text{Excess } CH_3I} RCH_2CH_2\overset{+}{N}(CH_3)_3 \ I^- \xrightarrow{\text{Ag}_2O}$$

$$RCH_2CH_2\overset{+}{N}(CH_3)_3 \ OH^- \xrightarrow{\text{Heat}} RCH=CH_2 + (CH_3)_3N + H_2O$$

The hydrogenation of acetylenes described in Chapter 12 is very important because it is possible in this way to prepare an alkene in a stereospecific fashion:

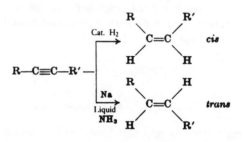

The Carbon–Carbon Triple Bond

Alkynes are normally prepared by the reaction of a sodium alkynide with an alkyl halide in liquid ammonia, as described in Chapter 12. The extension of this reaction to give a disubstituted acetylide is also described there:

$$RBr + Na^+ \ \overset{-}{C}{\equiv}CH \xrightarrow{\text{Liquid } NH_3} RC{\equiv}CH + Na^+Br^- \xrightarrow[\text{Liquid } NH_3]{\text{NaNH}_2}$$

$$RC{\equiv}\overset{-}{C} \ Na^+ \xrightarrow{\text{R'Br}} RC{\equiv}CR'$$

as is the formation of alkynols by the electron-donor addition of an acetylide anion to a ketone. Alkynes can also be prepared by elimination reactions although these are usually difficult to carry out. For instance, an alkene can be converted into an alkyne by first treating the alkene with a halogen to form a 1,2-dihalide and then treating this with a very powerful electron donor, such as sodamide:

$$RCH=CHR + Cl_2 \longrightarrow RCHClCHClR \xrightarrow{\text{NaNH}_2} RC{\equiv}CR$$

This reaction may give indifferent yields.

Chain-lengthening Sequences

We now finally wish to consider how we can take a sequence of reactions to build up or degrade a carbon chain. The simplest method of increasing the length of a carbon chain by one carbon atom at a time involves the use of cyanide:

$$ROH \xrightarrow{PBr_3} RBr \xrightarrow[C_2H_5OH]{KCN}$$

$$RCN \underset{LiAlH_4}{\overset{HOH}{\vert}} \begin{array}{l} RCO_2H \longrightarrow RCO_2C_2H_5 \longrightarrow RCH_2OH \\ \\ RCH_2NH_2 \end{array}$$

The alternative simple route is via the Grignard reagent and either carbon dioxide to give the carboxylic acid or methanal to give the alcohol:

$$ROH \xrightarrow{PBr_3} RBr \xrightarrow[(C_2H_5)O]{Mg} RMgBr \underset{CH_2O}{\overset{CO_2}{\vert}} \begin{array}{l} RCO_2H \\ \\ RCH_2OH \end{array}$$

Lengthening of the carbon chain by two or more carbon atoms can be achieved by using sodium acetylide:

$$ROH \xrightarrow{PBr_3} RBr \xrightarrow{NaC \equiv CH} RC \equiv CH \underset{\substack{Hg^{2+} \\ H_2SO_4}}{\overset{Na/NH_3(l)}{\vert}} \begin{array}{l} RC \equiv CNa \xrightarrow{RBr} RC \equiv CR' \\ \\ RCOCH_3 \end{array}$$

An alkyl chain can be attached to an aromatic nucleus by means of the Friedel–Crafts reaction.

A useful method of coupling of two alkyl groups involves the reaction of lithium dialkyl copper R_2LiCu with an alkyl halide $R'X$, in which an alkyl lithium reacts with copper (I) iodide and then subsequently another alkyl halide:

$$R\text{—}Br + Li \rightarrow RLi + LiBr$$
$$2RLi + CuI \rightarrow R_2LiCu + LiI$$
$$R_2CuLi + R'X \rightarrow R\text{—}R' + LiR + CuX$$

The lithium dialkyl copper is thought to react as $R_2Cu^-Li^+$.

Normally unreactive alkenyl halides, particularly alkenyl iodides, can be alkylated with this reagent with retention of the geometry of the alkene:

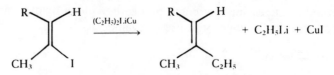

With the exception of the last two reactions, all the reactions we have discussed lead to straight chains. Branched chains can be obtained by the reaction of ketones with hydrogen cyanide:

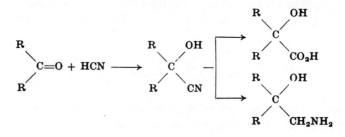

or by the reactions of carbonyl compounds with Grignard reagents:

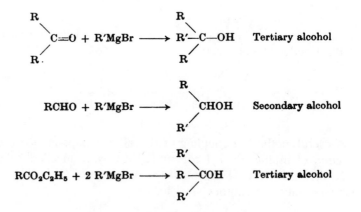

Shortening of the Carbon Chain

One of the best methods for degrading the carbon chain one carbon atom at a time is the so-called Barbier–Wieland degradation which simply consists of treating the ester of a carboxylic acid with a Grignard reagent, preferably phenylmagnesium bromide:

$$RCH_2CO_2C_2H_5 + 2\ C_6H_5MgBr \longrightarrow RCH_2\underset{\underset{C_6H_5}{|}}{\overset{\overset{C_6H_5}{|}}{C}}{-}OH \xrightarrow{-H_2O}$$

$$RCH{=}C\overset{C_6H_5}{\underset{C_6H_5}{\diagdown}}\ \xrightarrow[CrO_3]{O_3}\ RCO_2H + (C_6H_5)_2CO$$
$$\text{Benzophenone}$$

The acid produced in this reaction can be separated from the neutral benzophenone by dissolving it in sodium bicarbonate solution from which the free acid can be recovered. It can then be reesterified and the same reaction repeated. Stepwise degradation of carbon chains by this reaction was important in the past in studying the structures of natural products, but now spectroscopic methods using tiny quantities of material are used exclusively. If the chain terminates in an amino group then it is convenient to begin the series of degradations with a Hofmann elimination:

$$RCH_2CH_2NH_2 \xrightarrow[CH_3I]{Excess} RCH_2CH_2\overset{+}{N}(CH_3)_3I^- \longrightarrow$$

$$RCH_2CH_2\overset{+}{N}(CH_3)_3OH^- \xrightarrow{Heat} RCH{=}CH_2 \xrightarrow{O_3} RCHO \longrightarrow$$
$$RCO_2H$$

Many textbooks describe the decarboxylation of a carboxylic acid both as a method of preparing alkanes and as a method of degrading the carbon chain. Decarboxylation proceeds satisfactorily with ethanoic acid; sodium ethanoate treated with soda lime and heated does yield methane, and benzoic acid will decarboxylate readily to yield benzene. In general, however, this is not a satisfactory reaction and usually all sorts of other products are formed as well.

Note to the Student

We have endeavoured throughout the previous chapters to describe the reactions of different types of groups attached to a carbon chain and consider how they react and what makes them react in the way they do. In this last chapter we have listed nearly fifty sets of equations, but heaven forbid that any student should sit down and try and memorize these by rote! Of the fifty equations, only three are reactions which have not been described before. Our purpose in writing out again these reactions described in the previous chapters is simply to give you an opportunity of thinking about these reactions in a different light. Previously we were concerned with how and why the reactions occurred. Now we have looked at the reactions again with a view to seeing what practical application we could put them to.Many of the reactions described in this chapter you will have felt you were already familiar with, but many more you have probably forgotten. At the end of most chapters we suggested problems which usually involved thinking of reactions in this utilitarian or synthetic light. It is only by doing problems of this kind that a student can hope to become familiar with the vast number of reactions which organic compounds undergo. It is our hope, however, that having worked through the book thus far, you will feel you understand what is happening and why it is happening and are not just memorizing a lot of dull dry facts. By becoming familiar with how and why the reactions occur you should find it possible to devise sensible synthetic sequences without having to memorize endless lists of reactions. At the end of this chapter are two charts in which some of the principal reactions are represented diagrammatically, illustrating how different compounds can be interconverted. First attempt some of the problems with this chart in front of you and then when you become more familiar with the reactions cover up the chart and do the problems without any help.

The study of the chemistry of the compounds of carbon, involving as it does, on the one hand, a rapidly developing sector of industry and, on the other hand, the study of life itself is one of the most thrilling branches of science today. In this book we have been able to do little more than introduce you to the ABC of organic chemistry. Learning the alphabet is a dull business at the best of times. It is only when we realize that the sequences of letters joined together spell out words that it becomes interesting. We have endeavoured here to make the alphabet of organic chemistry as interesting as possible. It is up to you to now go on to learn to spell out the words, and finally make the words into sentences.

Problems

1. Starting from ethanol (C_2H_5OH), how would you prepare the following?
(a) Bromoethane (C_2H_5Br)   (b) Ethylamine ($C_2H_5NH_2$)
(c) Diethyl ether ((C_2H_5)$_2$O)   (d) Propan-1-ol (C_3H_7OH)
(e) Ethyl propanoate ($C_2H_5CO_2C_2H_5$)

2. Starting from benzene (C_6H_6) how would you prepare the following?
(a) Bromobenzene (C_6H_5Br)   (b) Nitrobenzene ($C_6H_5NO_2$)
(c) Aniline ($C_6H_5NH_2$)
(d) Benzonitrile (phenyl cyanide) (C_6H_5CH)

3. A liquid X ($C_6H_{13}N$) was treated with excess methyl iodide followed by moist silver oxide to yield an ionic substance Y ($C_9H_{21}NO$). On heating, Y yielded trimethylamine, $(CH_3)_3N$, water, and $Z(C_6H_{10})$. Z reacted rapidly with bromine (to yield a compound $C_6H_{10}Br_2$), with osmium tetroxide (to yield on hydrolysis $C_6H_{12}O_2$) and with ozone to yield on hydrolysis of the ozonide a dialdehyde which could be further oxidized to yield adipic acid, $HO_2C(CH_2)_4CO_2H$. Deduce the structures of X, Y and Z and elucidate the various reactions.

4. Starting from butan-1-ol ($CH_3CH_2CH_2CH_2OH$) how would you prepare the following compounds:

(a) $CH_3(CH_2)_4NH_2$ (b) $CH_3(CH_2)_2CH(CH_3)OH$

(c) $CH_3CH_2CHOHCH_2OH$ (d) $CH_3CH_2CO_2H$

(e) $CH_3(CH_2)_3C\equiv C(CH_2)_3CH_3$

5. How do electron donors react with a carbonyl bond? Starting from propanone how could the following be prepared?

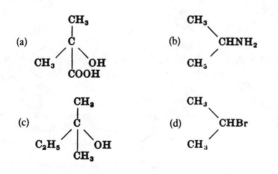

The Main Synthetic Sequences of Organic Compounds

Aliphatic Compounds

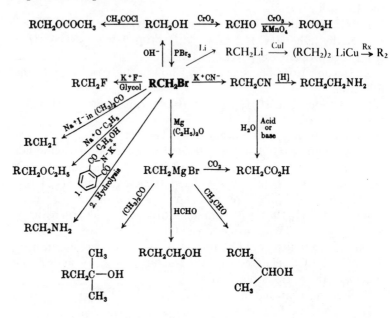

Aromatic Compounds

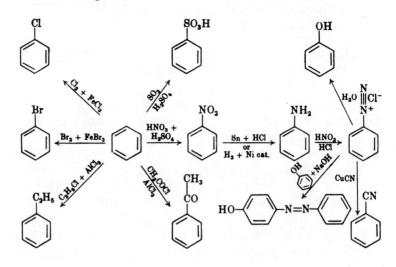

CHAPTER 20 ———————————————

Orbital Theory

Atomic Orbitals

At the beginning of Chapter 1 we said we would use Lewis' theory of valency, according to which chemical bonds were formed by atoms sharing electrons to form electron pairs, each atom acquiring a shell of electrons corresponding to the inert gas nearest to it in the periodic table. Thus the lithium atom loses an electron to form a lithium cation, Li^+, and fluorine gains an electron to form a fluoride anion, F^-. In most organic molecules carbon atoms share two, four or six electrons in electron-pair bonds in which the electrons in the bond are associated with the two atoms in the bond. We could depict a methane molecule, which consists of a carbon atom surrounded by four hydrogen atoms, as a tetrahedron with a hydrogen atom at each apex and a carbon atom in the centre. The 'bonds' are usually depicted as straight lines (cf. Chapter 1 for the representation of methane). This concept of electron-pair bonds was extended to double bonds (Chapter 6) and triple bonds (Chapter 12). However, we came across molecules and ions for which we could not draw a single electronic structure. In Chapter 9 we introduced resonance theory and subsequently made use of the concept of a resonance hybrid in Chapters 14 and 15. For the major part of organic chemistry, resonance theory is perfectly adequate. However, recently an alternative way of describing organic reactions has been developed called *pictorial orbital theory*. Both resonance theory and pictorial orbital theory have their origins in quantum mechanics.

The basis of all current theories of valency is the observation that atoms (and molecules) can only absorb fixed amounts of energy. A hydrogen atom can absorb light at particular frequencies (i.e. one atom can absorb a packet of energy, called a photon) and an excited atom can return to its ground state by emission of the photon light. The important point is that a hydrogen atom will only absorb and emit light at fixed energies; light with photons of greater or less energy are not absorbed. Thus a hydrogen atom will absorb light at a series of specific wavelengths, just as a stretched string will only vibrate at a series of specific frequencies.

Quantum mechanics tell us the energy and the most likely place to find an electron in an atom. The theory involves a quantity called the wave function ψ which has no direct significance itself, but ψ^2 defines the probability of finding an electron in an area in space. Thus in the ground state of the hydrogen atom the probability of finding inside a particular small volume of space (ψ^2) decreases as the distance from the nucleus increases. We can depict a sphere (a circle on the flat page) inside which there is, say, a 90 per cent probability of finding the electron. We call such a solution of the wave equation, which gives us the energy and the probability distribution, an 'orbital'. The lowest energy orbital is given the symbol s. The next higher energy level gives four spatial distributions (the plus and minus represent the sign of the wave function):

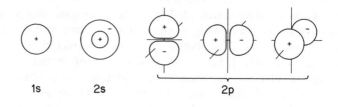

In the hydrogen atom the $2s$ and the $2p$ orbitals are of the same energy (degenerate), but for all other atoms the $2s$ orbital is of lower energy than the $2p$ orbitals. Notice there are 'nodes' (places where ψ is zero). Each orbital can take two electrons and no more. The building up of the periodic table is outside the scope of this chapter and we will ignore the higher energy atomic orbitals present in atoms beyond the second row of the periodic table.

Molecular Orbitals

If we can have electrons associated with individual atoms it requires no great extension of our theory to have molecular orbitals in which an electron (or electrons) is (are) associated with two or more atoms. The simplest example is the hydrogen molecule. If we bring two hydrogen atoms close together two molecular orbitals are formed: one of lower energy than the isolated atoms (bonding σ) and one of higher energy (antibonding σ^*). Although the wave function ψ has no physical interpretation (ψ^2 determines the electron density) the sign of the wave function is very important. When two hydrogen atoms approach each other they will interact to form a bond if the wave functions have the same sign; if they have opposite signs the state is repulsive and the two atoms will not combine:

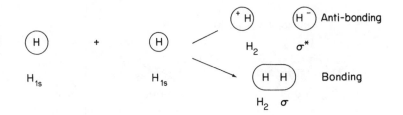

Alkanes are relatively inert compounds and are unaffected by ionic reagents. Alkyl halides have polar bonds which undergo displacement reactions (see Chapter 4). A typical replacement reaction would be as follows:

We can depict this kind of reaction in terms of frontier orbitals (i.e. the highest occupied molecular orbital, HOMO, interacting with the lowest unoccupied molecular orbital, LUMO).

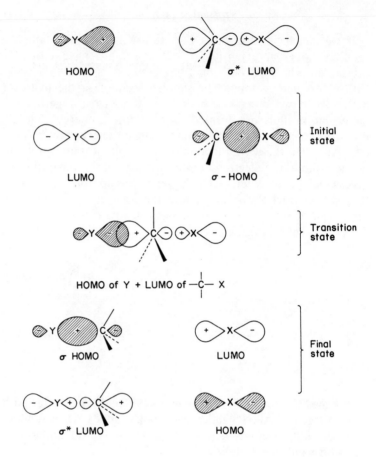

The carbon double bond we described in terms of two carbon atoms sharing the same four electrons:

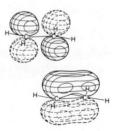

However, we made it clear that the two 'bonds' were not identical. In fact the highest occupied molecular orbital is of higher energy (hence more reactive) than a carbon–carbon single bond. Most reactions of alkenes involve electron acceptors attacking the occupied π orbitals. See Chapter 7 for the addition of electron acceptors to ethene.

When we come to the carbonyl bond we have two additional complications: firstly, oxygen is more electronegative than carbon and secondly, in simple Lewis terms the oxygen has two pairs of non-bonded electrons:

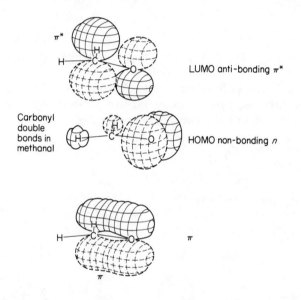

As the pictures show, an electron acceptor (for example H^+) would be expected to attack the HOMO on the oxygen, while an electron donor (for example CN^-) would be expected to attack the LUMO at the carbon.

So far we have been concerned with molecular orbitals which encompass two atoms. However, it is in molecules which have electrons delocalized over three or more atoms that pictorial orbital theory comes into its own. In Chapter 9 we found that we could not draw unique electron-pair structures for the ethanoate ion. Instead we described it in terms of two resonance hybrids. In pictorial orbital theory there are three delocalized π orbitals, one bonding, one non-bonding and one antibonding, which we can neglect. The lowest energy orbital (ψ_1) is delocalized over the whole carboxyl group; if we add two more electrons to the non-bonding orbital (ψ_2) these will appear at the two oxygen atoms with a node at the centre carbon atom:

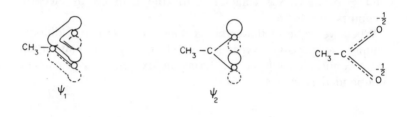

The carboxylate anion carries a negative charge and will be shared between the two oxygen atoms.

Buta-1,3-diene has four mobile electrons spread over four molecular orbitals, two bonding and two antibonding; the four 'π electrons' occupy the two bonding orbitals:

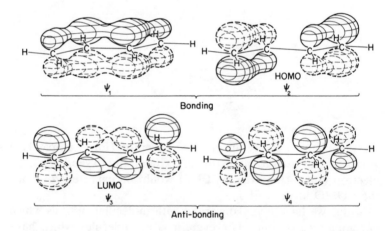

However, the principal example of delocalized electrons is benzene and its derivatives. There are six atoms contributing to the delocalized system which lead to six molecular orbitals, three

bonding and three antibonding; each carbon atom contributes one
electron to the delocalized system:

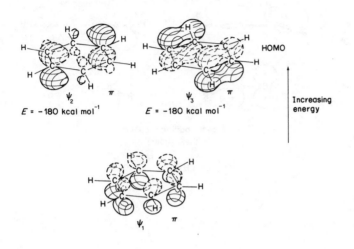

$E = -180$ kcal mol^{-1} $\quad\quad E = -180$ kcal mol^{-1} $\quad$ HOMO

Increasing
energy

ψ_2 $\quad\pi$ $\quad\quad\psi_3$ $\quad\pi$

ψ_1 $\quad\pi$

The highest and the lowest energy orbitals are unique, but the
four intermediate orbitals occur in degenerate (i.e. of the same
energy) pairs.

There is a class of reaction the mechanism of which is hard to
understand in terms of simple electron-pair transfer, in particular
the Diels–Alder reaction we have encountered in previous chapters (7 and 14). In these reactions there is no net transfer of
electrons. In the transition state the π electrons appeared to be
delocalized in a circle and the product is a cyclic molecule:

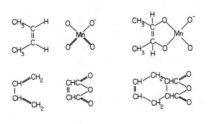

These cyclic addition reactions can be depicted as the simultaneous transfer of electrons to and from the reacting molecules:

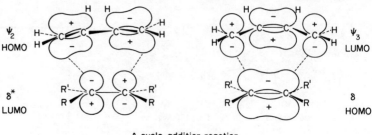

A cyclo-addition reaction
(Dielo-Alder)

ANSWERS TO PROBLEMS

Chapter 1

1. Isomers of C_7H_{16}:

    1. Isomers of C_7H_{16}:

$CH_3CH_2CH_2CH_2CH_2CH_2CH_3$ Heptane

$CH_3CH_2CH_2CH_2CHCH_3$ 2-Methylhexane
                  $|$
                CH_3

$CH_3CH_2CH_2CHCH_2CH_3$ 3-Methylhexane
               $|$
             CH_3

$CH_3CH_2CHCH_2CH_2$ 3-Ethylpentane
           $|$
        CH_2CH_3

           CH_3
           $|$
$CH_3CH_2CH_2CCH_3$ 2,2-Dimethylpentane
           $|$
           CH_3

CH_3CH_2CH—$CHCH_3$ 2,3-Dimethylpentane
           $|$ $|$
         CH_3 CH_3

$CH_3CHCH_2CHCH_3$ 2,4-Dimethylpentane
      $|$ $|$
    CH_3 CH_3

          CH_3
          $|$
$CH_3CH_2CCH_2CH_3$ 3,3-Dimethylpentane
          $|$
          CH_3

     CH_3 CH_3
     $|$ $|$
CH_3CH—C—CH_3 2,2,3-Trimethylbutane
           $|$
           CH_3

2. Possible conformations of methylcyclohexane:

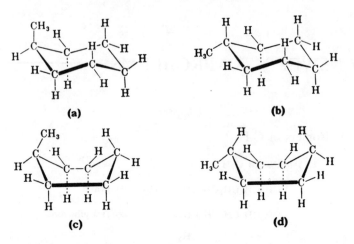

(a)   (b)

(c)   (d)

These four structures represent conformations of methyl-cyclohexane in which the tetrahedral arrangement of the carbon atoms is maintained. These are not *isomers*, but different arrangements which this molecule can assume. At normal temperatures the molecule is changing very rapidly from one form to the other. At any one moment the most probable conformation is b followed by a; d and c are much less probable, and represent the extreme conformations of a 'twist' form.

Chapter 3

1. If iodination of methane were to take place it would involve the same steps as chlorination:

$$I_2 + h\nu \xrightarrow{\ 1\ } 2\,I\cdot \qquad \text{Initiation}$$

$$\left. \begin{array}{l} CH_4 + I\cdot \xrightarrow{\ 2\ } CH_3\dot{} + HI \\[4pt] CH_3\dot{} + I_2 \xrightarrow{\ 3\ } CH_3I + I\cdot \end{array} \right\} \text{Propagation}$$

$$\left. \begin{array}{l} I\cdot + I\cdot + M \xrightarrow{\ 4\ } I_2 + M \\[4pt] CH_3\dot{} + I\cdot \xrightarrow{\ 5\ } CH_3I \\[4pt] CH_3\dot{} + CH_3\dot{} \xrightarrow{\ 6\ } C_2H_6 \end{array} \right\} \text{Termination}$$

Taking the values of the bond-dissociation energies given in the question and the value $D(CH_3\!-\!H) = 427$ kJ mol^{-1} we have:

$$\Delta H_2 = + 427 - 297 = + 130 \text{ kJ mol}^{-1}$$
$$\Delta H_3 = + 151 - 210 = - 59 \text{ kJ mol}^{-1}$$

Reaction 2 is endothermic to the extent of 130 kJ mol^{-1}. It is therefore a most unfavourable reaction and iodination of methane is not practicable.

2. We have seen that the bromination (at 150°C) at the secondary position in propane occurs 90 times faster than attack at the primary position. There are four hydrogen atoms attached to secondary carbon atoms and six attached to primary carbon atoms in butane. Therefore if we brominate butane at 150°C we would expect to obtain 2-bromobutane and 1-bromobutane in the proportions 60:1. Hydrogen atoms attached to a tertiary carbon atom are the least strongly bound. We therefore expect bromination at a tertiary site to occur considerably faster than a secondary site. In 2-methylpropane there is one tertiary site, the remainder being unreactive primary sites. Thus there is one site very much more reactive than all the others and we would expect bromination of 2-methyl(propane) to yield 2-bromo-2-methylpropane almost exclusively, which it does.

Fluorination of propane at room temperature favours attack at the secondary position by a factor of 1.5:1. Fluorination of butane under the same conditions yields equal amounts of 1-fluoro- and 2-fluorobutane. Fluorination is so much less selective that, in spite of the fact that a hydrogen atom bonded to a tertiary carbon atom is appreciably less tightly bound than a hydrogen atom bonded to a primary carbon, fluorination of 2-methylpropane yields more 1-fluoro-2-methylpropane than 2-fluoro-2-methylpropane.

Chapter 4

1. In bromoethane the bromine atom is more electronegative than carbon, and the C—Br bond is polar:

$$\overset{\delta+}{\text{C}}—\overset{\delta-}{\text{Br}}$$

It will react with electron donors, e.g. anions (a), (d) and (h):

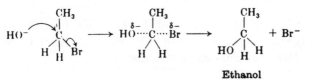

Ethanol

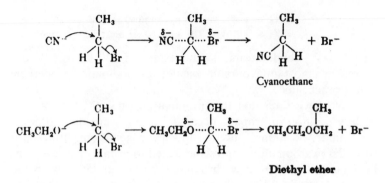

Cyanoethane

Diethyl ether

With the halogen anions (g) and (i) the reaction is similar but reversible:

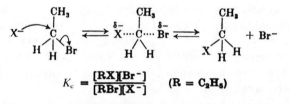

$$K_c = \frac{[RX][Br^-]}{[RBr][X^-]} \quad (R = C_2H_5)$$

For fluoride K_c is large, and provided the reaction is carried out in a solvent in which F^- is only weakly solvated (e.g. glycol) the reaction goes almost to completion. For iodide $K_0 < 1$, but NaI is soluble in acetone and NaBr is not, so that by carrying out the reaction in acetone it may be forced to completion.

Ammonia (e) carries no negative charge, but it does possess a pair of electrons which are readily donated:

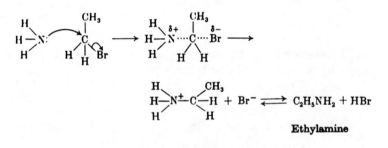

Ethylamine

Ethylamine will then react with further ethyl bromide to yield diethylamine, and this in turn will give triethylamine and ultimately tetraethylammonium bromide.

A chlorine atom (c) will react with bromoethane in the same way as it will react with ethane, but note that there are two possible products:

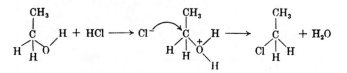

Bromoethane does not react with electron acceptors, i.e. cations (b) or nitric acid (f) which ionizes as $H_3O^+ + NO_3^-$ in aqueous solution.

2. The double molecular ion (M^+ 94 and 96) of equal abundance indicates the presence of bromine (isotopic natural abundance ^{79}Br: ^{81}Br::1:1). The difference, 15 (96−81 or 94−79), indicates CH_3.

In conclusion, the compound is bromomethane, CH_3Br. Note the peaks at m/e 79 and 81, Br^+ and at 15, CH_3^+.

Chapter 5

1. (1) (a) Dry hydrogen chloride will react with anhydrous alcohol to yield chloroethane and water:

$$
\begin{array}{c}
CH_3 \\
| \\
\underset{\underset{H}{|}}{\overset{}{C}} \overset{}{\underset{O}{}} H + HCl \longrightarrow Cl^- \\
H
\end{array}
\quad
\begin{array}{c}
CH_3 \\
| \\
\underset{\underset{H}{|}}{\overset{}{C}} \overset{}{\underset{\overset{+}{O}}{}} H \\
H \quad H
\end{array}
\longrightarrow
\begin{array}{c}
CH_3 \\
| \\
\underset{\underset{H}{|}}{\overset{}{C}} \\
Cl \quad H
\end{array}
+ H_2O
$$

Note that ethanol will not react with hydrochloric acid (aqueous solutions of hydrogen chloride) and nor will a solution of sodium chloride react with ethanol; i.e.

$$Cl^- + C_2H_5{-}OH \not\longrightarrow C_2H_5Cl + OH^-$$

does not occur (see second paragraph of Chapter 5).

(b) $C_2H_5OH + Na \longrightarrow C_2H_5O^-Na^+ + \frac{1}{2}H_2$

(c) No reaction.

(d) $C_2H_5OH + Na^+NH_2^- \longrightarrow C_2H_5O^- + Na^+ + NH_3$

(e) No reaction.

2 (a) $C_2H_5NH_2 + HCl \longrightarrow C_2H_5\overset{+}{N}H_3Cl^-$

(b) $C_2H_5NH_2 + Na \xrightarrow{\text{Slow}} C_2H_5NH^-Na^+ + \frac{1}{2}H_2$

(c) $C_2H_5NH_2 + H_2O \rightleftarrows C_2H_5NH_3^+ + OH^-$ $K_b = 4 \times 10^{-4}$

(d) No reaction. (Ethylamine is a stronger base than ammonia.)

(Carrying out these reactions would be complicated by the fact that ethylamine is a gas at room temperature.)

(e)

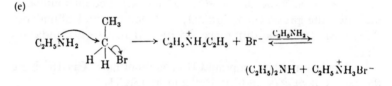

$$(C_2H_5)_2\overset{..}{N}H + C_2H_5NH_3^+Br^-$$

$$(C_2H_5)_3N + C_2H_5\overset{+}{N}H_3Br^-$$

2. (a)

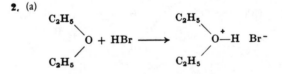

Rewriting the diethylhydroxonium ion, we get:

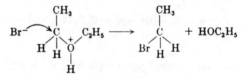

Note that this cleaveage of ethers by hydrogen halides will only occur when the anhydrous hydrogen halide is used. Dilute aqueous solutions of hydrogen bromide (i.e. hydrobromic acid) would not react.

(b) $Cl_2 + (C_2H_5)_2O \longrightarrow$ no reaction in the dark

$$Cl_2 + h\nu \longrightarrow 2\ Cl\cdot \qquad \text{Initiation}$$

$$Cl\cdot + CH_3CH_2OC_2H_5 \longrightarrow \dot{C}H_2CH_2OC_2H_5$$

$$Cl\cdot + CH_3CH_2OC_2H_5 \longrightarrow CH_3\dot{C}HOC_2H_5$$

$$CH_3\dot{C}HOC_2H_5\ (\text{or}\ \dot{C}H_2CH_2OC_2H_5) + Cl_2 \longrightarrow$$
$$CH_3CHClOC_2H_5\ (\text{or}\ CH_2ClCH_2OC_2H_5) + Cl\cdot$$

Chain propagation

There will be the usual termination steps, e.g.

$$Cl\cdot + Cl\cdot + M \longrightarrow Cl_2 + M$$
$$R\cdot + R\cdot \longrightarrow R_2$$
$$R\cdot + R'\cdot \longrightarrow R-R'$$
$$R'\cdot + R'\cdot \longrightarrow R'_2$$
$$(R'\cdot)\ R\cdot + Cl\cdot \longrightarrow RCl\ (R'Cl)$$

In practice the chlorination of ethers at low temperatures in the liquid phase yields predominantly the α-chloro ether (i.e. $CH_3CHClOC_2H_5$). In the gas phase the reaction is much more complicated and outside the scope of this book.

(c) No reaction.

3. The infrared spectrum shows the presence of a C—O bond, by the broad absorption at 1100 cm^{-1} and the absence of any other functional group such as OH or NH.

The ^1H n.m.r. spectrum shows the triplet, quartet in the ratio 3:2, characteristic of an ethyl group with the CH_2 group deshielded to indicate the presence of an adjacent heteroatom such as oxygen.

The ^{13}C n.m.r. spectrum serves to confirm the presence of the ethyl group; two types of carbon, one more deshielded than the other.

Mass spectrum. Molecular weight 74. From the other spectral data the structural unit present is C_2H_5O. M$^+$ – oxygen: $74-16 = 58 = 2 \times 29 = (C_2H_5)_2$.

The compound is diethyl ether $(C_2H_5)_2O$.

Chapter 6

1. Alkenes with the molecular formula C_5H_{10}:

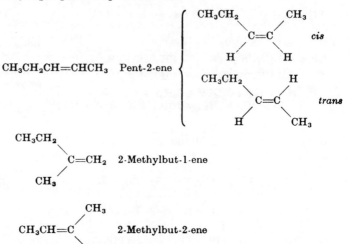

$CH_3CH_2CH_2CH{=}CH_2$ Pent-1-ene

$CH_3CH_2CH{=}CHCH_3$ Pent-2-ene

2-Methylbut-1-ene

2-Methylbut-2-ene

3-Methylbut-1-ene

2. (a) To ensure elimination from bromoethane, a primary alkyl halide, it is necessary to use a very strong base; e.g.

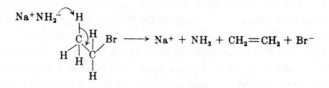

(b) 2-Bromopropane is a secondary alkyl halide and thus undergoes elimination fairly readily, so it is not necessary to use such a strong base:

$$\begin{array}{c} CH_3 \\ \diagdown \\ CHBr + Na^+ \ OC_2H_5^- \ \xrightarrow{C_2H_5OH} \\ \diagup \\ CH_3 \end{array} \qquad CH_3CH{=}CH_2 + Na^+Br^- + C_2H_5OH$$

(c) Methyl 2-propyl ether (2-methoxypropane) cannot be prepared by the reaction of 2-bromopropane with sodium methoxide (we have just used sodium ethoxide to promote elimination). Very careful treatment with an aqueous alkaline solution will result in some elimination, but the predominant reaction will be displacement:

Treatment of the propan-2-ol as formed with sodium metal will yield the corresponding alkoxide which we may then react with iodomethane.

$$CH_3CHOHCH_3 + Na \longrightarrow (CH_3)_2CHO^- + Na^+ + \tfrac{1}{2} H_2$$

3. The infrared spectrum shows the presence of a $={=}C{-}H$ group ($3100 \ cm^{-1}$), a $C{=}C$ bond ($1650 \ cm^{-1}$) and $={=}CH_2$ (bending CH vibrations at 1000 and $900 \ cm^{-1}$).

The ^1H n.m.r. spectrum indicates the presence of CH_3 groups ($\delta \ 1.0$ ppm) and vinyl hydrogens in the ratio 9:3, suggesting the presence of a $(CH_3)_3$ group attached to a carbon–carbon double bond.

The ^{13}C n.m.r. spectrum confirms the presence of a $(CH_3)_3C$ group; two types of 'alkyl' carbons at $\delta \ 24$ and 35 ppm and the carbon–carbon double bond carbons at $\delta \ 114$ and 144 ppm.

In conclusion, the difference between the molecular weight shown by the mass spectrum (84) and a $(CH_3)_3C$ group (57) is 27 or $HC{=}CH_2$, leading to the structure $(CH_3)_3CCH{=}CH_2$ or 3,3-dimethylbut-1-ene.

Chapter 7

1. (a)

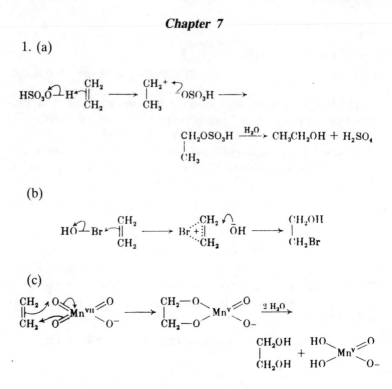

(b)

(c)

(*Note.* This represents a possible mechanism; the reactions in alkaline permanganate are extremely complex.)

(d)

(e)

$$R-(CH_2CH_2)_2 \xrightarrow{CH_2=CH_2} , \text{ etc.}$$

(R· = initiating radical)

2. Heterolytic addition
   (a) Attack by an electron acceptor

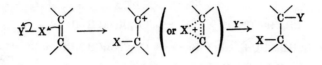

(a′) Attack by an electron acceptor

$$x \overset{\frown}{-} Y \rightleftharpoons X^+ + Y^-$$

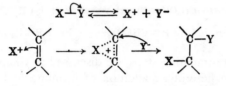

(b) Attack by an electron donor (uncommon in hydrocarbon olefins):

$$x \overset{\frown}{-} Y \rightleftharpoons X^- + Y^+$$

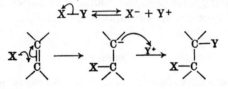

Homolytic attack:

$$\overset{\frown}{X} \overset{\frown}{Y} \dashrightarrow X \cdot + Y \cdot$$

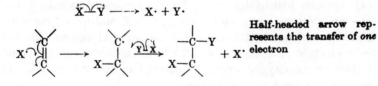

Half-headed arrow represents the transfer of *one* electron

Heterolytic addition of bromine to but-2-ene (in solution in the dark):

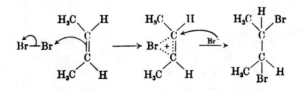

Homolytic addition of bromine to but-2-ene (in the gas phase in the light):

$$Br_2 \xrightarrow{hv} 2\ Br\cdot$$

$$CH_3CH=CHCH_3 \quad Br\cdot \longrightarrow CH_3CHBrCHCH_3$$

$$CH_3CHBr\dot{C}HCH_3 \quad Br-Br \longrightarrow CH_3CHBrCHBrCH_3 + Br\cdot \quad \text{Chain reaction}$$

Under these conditions hydrogen abstraction is also possible:

$$Br\cdot \quad H-CH_2CH=CHCH_3 \longrightarrow BrH + CH_2CH=CHCH_3$$

$$Br-Br \quad \dot{C}H_2CH=CHCH_3 \longrightarrow Br\cdot + BrCH_2CH=CHCH_3$$

The chain-terminating steps are discussed in Chapter 3.

From the heterolytic addition of bromine to but-2-ene, only 2,3-dibromobutane can be formed. In the homolytic reaction, in addition to 2,3-dibromobutane, 1-bromobut-2-ene and 1,2,3-tribromobutane may also be found (the latter as a result of addition of bromine to 1-bromobut-2-ene).

Chapter 8

1. This question is to illustrate the difference in reactivity of the carbon–carbon and carbon–oxygen double bonds. Three of the reagents are electron donors: (1) aqueous sodium cyanide, (4) aqueous hydroxylamine and (5) aqueous ammonia. These reagents will not react with ethene but will add to the carbonyl bond of methanal attacking the carbon atom. Hydrogen chloride is an electron acceptor which will add to ethene and to methanal (attacking the oxygen atom). Aqueous sodium chloride will add neither to ethene nor to methanal.

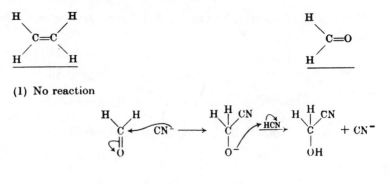

(1) No reaction

(2) No reaction　　　　　　　　　　　　No reaction

(3)

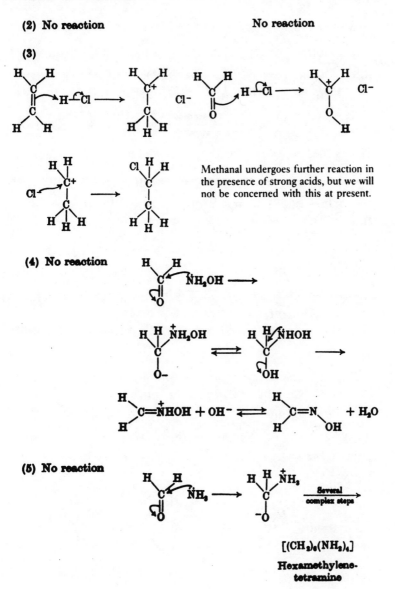

Methanal undergoes further reaction in the presence of strong acids, but we will not be concerned with this at present.

(4) No reaction

(5) No reaction

$[(CH_2)_6(NH_2)_4]$

**Hexamethylene-
tetramine**

2. This question illustrates the similarity of the electron-donating reagents which react with alkyl halides by a substitution reaction with those which react with aldehydes and ketones by addition.

(1a)

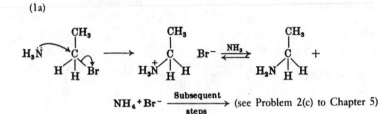

$$NH_4{}^+Br^- \xrightarrow[\text{steps}]{\text{Subsequent}} \text{(see Problem 2(c) to Chapter 5)}$$

(1b)

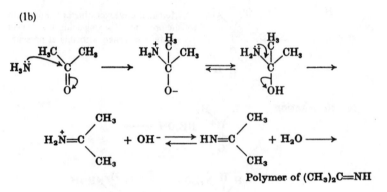

Polymer of $(CH_3)_2C=NH$

(2a)

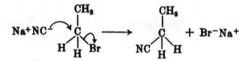

(2b)

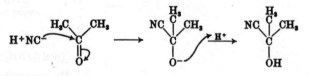

(Although both (2a) and (2b) involve attack by CN^-, (2a) goes to completion if bromoethane is treated with potassium cyanide in ethanol, whereas (2b) goes to completion in aqueous sodium or potassium cyanide solution to which 1 mole of acid has been added.)

(3a)

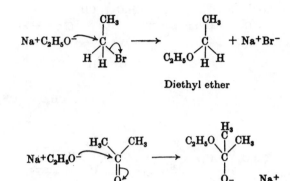

Diethyl ether

(3b)

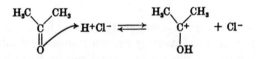

The product of this reaction is on acidification a hemi-acetal which is unstable in water and hydrolyses to regenerate propanone and ethanol.

(4a) No reaction.

(4b) Electron acceptors add to the carbonyl bond attacking the oxygen atom and leaving a carbocation:

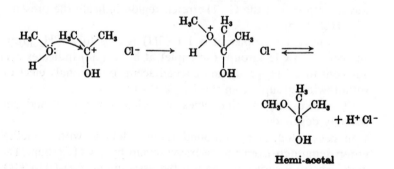

The carbocation ion is much more susceptible to attack by electron donors and weak electron donors such as methanol react readily.

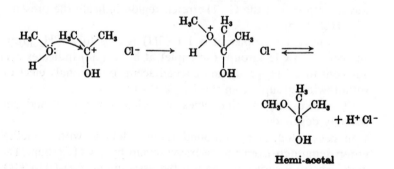

Hemi-acetal

The difference between this reaction and the base-catalysed addition (3b) is that this reaction goes further:

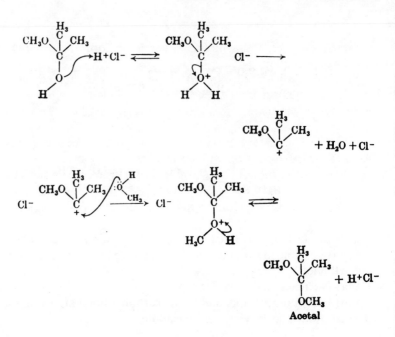

All these steps are reversible although we have only put double arrows for reactions involving proton transfer, because these are extremely rapid. Although the acetal can be hydrolysed back to propanone and two moles of ethanol this reaction is very much slower than the hydrolysis of the hemi-acetal.

3. Infrared. The absorptions at 2885 and 2740 cm^{-1} at the short-wavelength end of the C—H stretch region indicate the presence of a HC=O group.

^{1}H n.m.r. The triplet at δ 1.1 (3H) is due to a CH$_3$ group adjacent to a CH$_2$ group; the triplet at δ 9.8 (^{1}H) indicates a H adjacent to a CH$_2$ group and also adjacent to a strongly electron withdrawing group (from the i.r., a C=O).

^{13}C n.m.r. Shows three types of carbon—two alkyl and one strongly deshielded.

In conclusion, the compound is an aldehyde with a methyl group separated from the carbonyl group by a CH$_2$ group. The mass spectrum serves to confirm the assignment: CH$_3$CH$_2$CHO, propanal, molecular weight 58.

Chapter 9

1. Infrared. The absorption at 1740 cm^{-1} shows the presence of a carbonyl group.

^{1}H n.m.r. Shows the presence of a CH_3CH_2 group (triplet δ 1.25 3H and quartet δ 4.12, 2H) with the CH_2 group adjacent to an electron withdrawing atom (possibly oxygen) and another CH_3 group with no adjacent protons.

^{13}C n.m.r. Shows the presence of four types of carbon atom, one in a strongly electron withdrawing environment (C=O).

In conclusion, subtracting $CH_3CH_2O(45)$ from the molecular weight (88) leaves 43 or CH_3 C=O, leading to the structure $CH_3CH_2OOCCH_3$, ethyl ethanoate.

Chapter 10

1. (a) The reaction between ethanol and ethanoic acid is a reversible one, the equilibrium being established very slowly:

$$CH_3CO_2H + C_2H_5OH \rightleftharpoons CH_3CO_2C_2H_5 + H_2O$$

The reaction can be greatly accelerated by the addition of a strong mineral acid as catalyst:

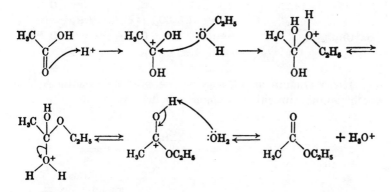

If, instead of adding just a trace of mineral acid as a catalyst, we add an excess of sulphuric acid then all the water will be converted into hydroxonium sulphate ($H_3\overset{+}{O}$ HSO_4^-) and so the reaction will go to completion.

(b) Ethanoamide is most conveniently prepared by the direct reaction of ethyl ethanoate with ammonia:

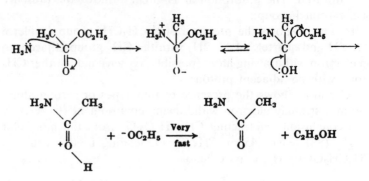

Alternatively, both ethyl ethanoate and ethanoamide may be prepared by the reaction of ethanoyl chloride (see below) with ethanol and ammonia respectively. Ethanoamide can also be prepared by heating ammonium ethanoate but this reaction is restricted to ethanoamide and is not general for other amides.

(c) Ethanoic anhydride can be prepared in the laboratory by treating ethanoyl chloride with sodium ethanoate, i.e.

$$CH_3CO_2H + SOCl_2 \longrightarrow CH_3COCl + HCl + SO_2 \text{ (cf. page 00)}$$

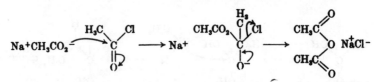

(d) N,N-Dimethylamide can be prepared by treating N,N-dimethylamine with either ethanoic anhydride or ethanoyl chloride:

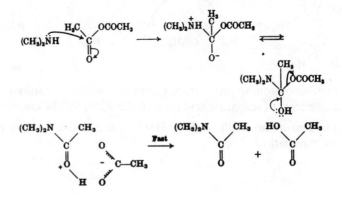

Chapter 11

1. The simplest laboratory test to distinguish between primary, secondary and tertiary amines is to treat the unknown substance with dilute aqueous nitrous acid. If the unknown base is a primary aliphatic amine, nitrogen will be evolved:

$$RCH_2NH_2 + HONO \longrightarrow [RCH_2NHNO] \longrightarrow [RCH_2N_2OH] \longrightarrow$$

$$[RCH_2N_2{}^+\,OH^-] \longrightarrow [RCH_2{}^+] + N_2\uparrow \xrightarrow{H_2O}$$

$$RCH_2OH + \text{other products}$$

The compounds in square brackets are transient intermediates, none of them capable of isolation.

If the unknown base is a secondary amine, a nitrosamine will be formed. Nitrosamines are yellow oils, insoluble in dilute aqueous acid if there are more than five carbon atoms. (If the secondary amine is diethylamine, for example, no oil will be precipitated but the solution will turn deep yellow.)

If the unknown base is a tertiary aliphatic amine, no reaction will occur and the free base may be recovered from the aqueous nitrous acid solution by treatment with aqueous sodium hydroxide.

Chapter 12

1. (a) Bromoethane is treated with a solution of potassium cyanide in ethanol (cf. Chapter 4):

$$C_2H_5Br + KCN \xrightarrow{C_2H_5OH} C_2H_5CN + KBr$$

Hydrolysis of cyanoethane, preferably using an acid catalyst:

$$C_2H_5CN + H_2O \xrightarrow{H^+A^-} C_2H_5CO_2H$$

(b)

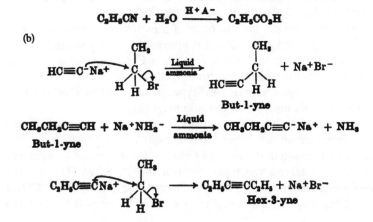

CH₃CH₂C≡CH + Na⁺NH₂⁻ $\xrightarrow[\text{ammonia}]{\text{Liquid}}$ CH₃CH₂C≡C⁻Na⁺ + NH₃
But-1-yne

This reaction may be carried out in two stages without separating the butyne. The first stage involves the preparation of sodamide in liquid ammonia. Ethyne gas is then passed through the liquid ammonia suspension of sodamide and sodium ethynide is formed. Alternatively, ethyne gas may be passed directly into a solution of sodium in liquid ammonia, but the reaction is rather vigorous and much ethene is formed:

$$C_2H_2 + Na \longrightarrow C_2HNa + \tfrac{1}{2} H_2; \quad C_2H_2 \xrightarrow{H_2} C_2H_4$$

Bromoethane is added to the solution of sodium ethynide in liquid ammonia and but-1-yne is formed. A fresh suspension of sodamide in liquid ammonia is made in a separate reaction vessel and added to the butyne–ammonia mixture. The butyne reacts rapidly with the sodamide to form the sodium butynide and the second portion of bromoethane is now added. The hex-3-yne is isolated by allowing most of the ammonia to evaporate and then adding water. The water dissolves the sodium bromide and leaves the hexyne floating on a dilute aqueous solution of ammonia.

(c) Treatment of hex-3-yne with mercuric sulphate in aqueous sulphuric acid:

$$C_2H_5C{\equiv}CC_2H_5 \xrightarrow[H_2O + H_2SO_4]{Hg^{2+}} C_3H_7COC_2H_5$$

(d) Hydrogenation of cyanoethane will yield 1-aminopropane:

$$C_2H_5C{\equiv}N \xrightarrow{H_2} C_3H_7NH_2$$

2. Infrared. Shows the presence of an 'acidic' C—H stretch at 3300 cm^{-1} and a triple bond stretch (2100 cm^{-1}).

^{1}H n.m.r. Shows the presence of a methyl group adjacent to a CH$_2$ group (triplet, 3H at δ 1.0 ppm); of the two multiplets, the one less shielded (δ 1.3–1.8 ppm) is a CH$_2$ group, the other (δ 1.9–2.3 ppm) is caused by three other protons.

^{13}C n.m.r. Shows three alkyl type carbons and two less shielded carbons (δ 68 and 84) in the alkyne region.

In conclusion, the ^{13}C spectrum indicates a 5-carbon chain and the presence of an alkyne triple bond. The presence of this functional group in the form of C≡C—H is supported by evidence from the infrared spectrum. Subtracting 25 (C≡CH) from the molecular weight leaves 43, which could be either CH$_3$—CO or CH$_3$CH$_2$CH$_2$. The absence of a carbonyl group (i.r.) and the

presence of a 3-carbon chain leads to the structure CH_3CH_2-
$CH_2C{\equiv}CH.$, pent-1-yne.

Chapter 13

1. All these reactions involve butylmagnesium bromide:

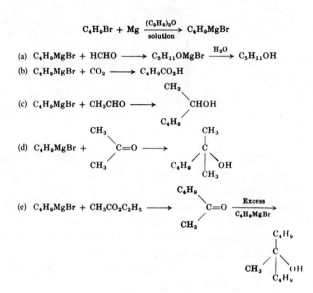

$$C_4H_9Br + Mg \xrightarrow[\text{solution}]{(C_2H_5)_2O} C_4H_9MgBr$$

(a) $C_4H_9MgBr + HCHO \longrightarrow C_5H_{11}OMgBr \xrightarrow{H_2O} C_5H_{11}OH$

(b) $C_4H_9MgBr + CO_2 \longrightarrow C_4H_9CO_2H$

(c) $C_4H_9MgBr + CH_3CHO \longrightarrow$ $\begin{array}{c} CH_3 \\ \diagdown \\ CHOH \\ C_4H_9 \diagup \end{array}$

(d) $C_4H_9MgBr +$ $\begin{array}{c} CH_3 \\ \diagdown \\ C{=}O \\ \diagup \\ CH_3 \end{array}$ $\longrightarrow$ $\begin{array}{c} CH_3 \\ | \\ C_4H_9\!-\!C\!-\!OH \\ | \\ CH_3 \end{array}$

(e) $C_4H_9MgBr + CH_3CO_2C_2H_5 \longrightarrow$ $\begin{array}{c} C_4H_9 \\ \diagdown \\ C{=}O \\ \diagup \\ CH_3 \end{array} \xrightarrow[C_4H_9MgBr]{\text{Excess}}$

$$\begin{array}{c} C_4H_9 \\ | \\ CH_3\!-\!C\!-\!OH \\ | \\ C_4H_9 \end{array}$$

Chapter 14

1. (a)

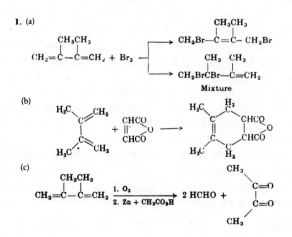

$$CH_2{=}\!\!\!\underset{\underset{CH_3}{|}}{C}\!\!\!-\!\!\!\underset{\underset{CH_3}{|}}{C}\!\!\!=\!\!CH_2 + Br_2$$

$$\longrightarrow CH_2Br\!-\!\!\!\underset{\underset{CH_3}{|}}{C}\!\!\!=\!\!\!\underset{\underset{CH_3}{|}}{C}\!\!\!-\!CH_2Br$$

$$\longrightarrow CH_2BrCBr\!-\!\!\!\underset{\underset{CH_3}{|}}{C}\!\!\!=\!\!CH_2$$

Mixture

(b)

$$\begin{array}{c} H_3C \diagdown \quad \diagup CH_2 \\ C \\ \| \\ C \\ H_3C \diagup \quad \diagdown CH_2 \end{array} + \begin{array}{c} CHCO \\ \| \quad \diagdown O \\ CHCO \end{array} \longrightarrow$$

(c)

$$CH_2{=}\!\!\!\underset{\underset{CH_3}{|}}{C}\!\!\!-\!\!\!\underset{\underset{CH_3}{|}}{C}\!\!\!=\!\!CH_2 \xrightarrow[\text{2. Zn} + CH_3CO_2H]{\text{1. } O_3} 2\,HCHO + \begin{array}{c} CH_3 \\ \diagup \\ C{=}O \\ C{=}O \\ \diagdown \\ CH_3 \end{array}$$

2.

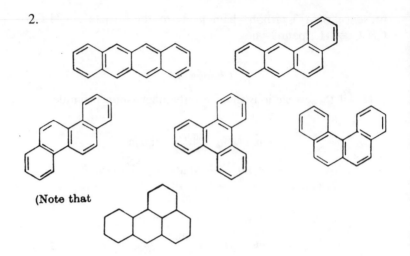

(Note that

will not be a stable molecule. Try drawing Kekulé, i.e. cyclohexa-triene, rings keeping each carbon atom surrounded by eight elec-trons in its outer shell.)

3. Absorptions at 1600 cm^{-1} (arene C=C) and at 750 and 700 cm^{-1}(monosubstituted arene) indicates an arene.

The ^{1}H n.m.r. shows the presence of alkyl protons and arene protons, the ratio 3:5 indicating a CH$_3$ and C$_6$H$_5$ group.

The ^{13}C n.m.r. shows the presence of five types of carbon, one alkyl and four alkene or arene.

CH$_3$ + C$_6$H$_5$ = 15 + 77 = 92; the molecular ion confirms that the compound is methyl benzene and the ^{13}C spectrum shows the presence of 2 *ortho*- + 2 *meta*- + 1 *para* carbon (126, 128, 129) and the carbon carrying the methyl group (δ 138 ppm).

4. The ^{1}H n.m.r. shows four arene protons, with a symmetrical absorption pattern, (centre at δ 6.8 ppm) characteristic of 1,4-disubstituted benzene, and two additional protons (δ 3.5 ppm) from the integration.

I.r. Double absorption at 3350–3550 cm^{-1} is characteristic of NH$_2$ (OH would be broad and single). The 1,4-substitution is confirmed by the weak absorption at 1800 cm^{-1} and the stronger absorption at 815 cm^{-1}.

The ^{13}C n.m.r. shows the presence of four types of carbon, all arene carbons, giving supoprt to the idea of 1,4-substitution.

MS. The pair of molecular ion peaks of equal abundance suggest the presence of bromine.

171–79 or 173–81 leaves 92
92–16 (NH_2) leaves 76 or C_6H_4

The compound is 1-amino-4-bromobenzene (*p*-bromoaniline):

Chapter 15

1.(a) $C_6H_5CH{=}CH{-}CH{=}CHC_6H_5 + Br_2 \longrightarrow$
   $C_6H_5CHBrCH{=}CHCHBrC_6H_5 + C_6H_5CHBrCHBrCH{=}CHC_6H_5$

(b)

This reaction is carried in the industrial preparation of Terylene. In industry *p*-xylene is mixed with oxygen and passed over vanadium pentoxide (see Chapter 18).

(c)

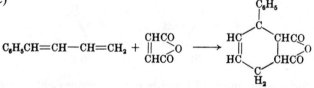

$C_6H_5CH{=}CH{-}CH{=}CH_2 + \begin{matrix} CHCO \\ \| \quad \rangle O \\ CHCO \end{matrix} \longrightarrow$

2. (a) $C_6H_5Br + Mg \xrightarrow[\text{solution}]{(C_2H_5)_2O} C_6H_5MgBr$

   $C_6H_5MgBr + CO_2 \longrightarrow C_6H_5CO_2H$

(b) $C_6H_6 + HNO_3 + H_2SO_4 \longrightarrow C_6H_5NO_2$
                                     Nitrobenzene

   $C_6H_5NO_2 + H_2 \xrightarrow[\text{or Sn + HCl}]{\text{Catalyst Ni}} C_6H_5NH_2$

(c)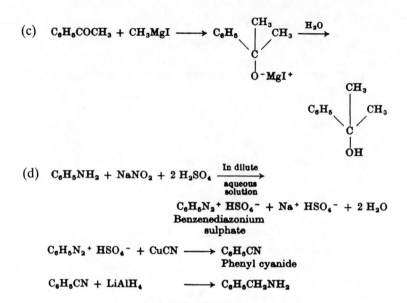

(d) $C_6H_5NH_2 + NaNO_2 + 2\ H_2SO_4 \xrightarrow[\text{aqueous solution}]{\text{In dilute}}$

$C_6H_5N_2{}^+\ HSO_4{}^- + Na^+\ HSO_4{}^- + 2\ H_2O$
Benzenediazonium sulphate

$C_6H_5N_2{}^+\ HSO_4{}^- + CuCN \longrightarrow C_6H_5CN$
Phenyl cyanide

$C_6H_5CN + LiAlH_4 \longrightarrow C_6H_5CH_2NH_2$

Notice that the 'double bonds' of the benzene nucleus do not take part in any of these reactions.

Chapter 16

1. (a)

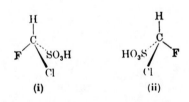

(i)   (ii)

ii is the mirror image of i and is not superimposable upon i.
    Similarly with (b) when the structure is written out more fully, i and ii are optically active.

(b)

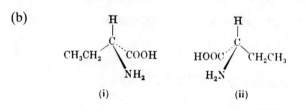

(i)   (ii)

(c)

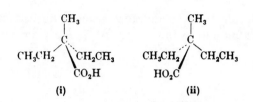

(i)   (ii)

This compound has two ethyl groups attached to the central carbon atom and ii can be made to coincide with i. The molecule is not asymmetric and does not show optical activity.

(d)

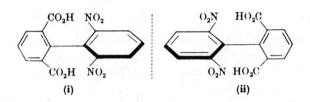

(i)   (ii)

ii, the mirror image of i, is superimposable upon i. The molecule is not optically active, even though the bulky carboxyl and nitro groups do not allow the two benzene rings to lie in the same plane.

(e)

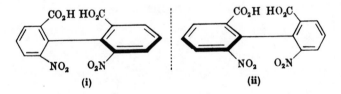

(i)   (ii)

The two mirror images are not superimposable in this case and the molecule is optically active.

(f)

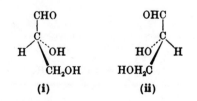

(i)   (ii)

The molecule has four different groups attached to the central carbon atom and, as in (a) and (b), it is asymmetric.

2.

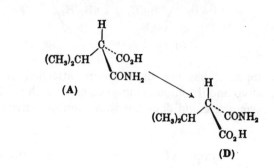

In the reaction sequence A→D the CO_2H and $CONH_2$ groups have been interchanged. This exchange has given us the mirror image of A, i.e. D could be written

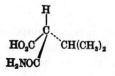

and will have a specific rotation of $-49°40'$.

Chapter 17

1. (a)

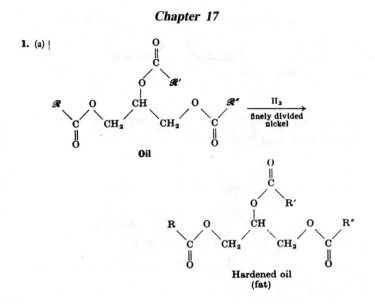

$\mathcal{R}, \mathcal{R}', \mathcal{R}''$ are hydrocarbon chains containing 15, 17 and 19 carbon atoms and one or two, possibly three, double bonds per chain. R, R' and R'' are hydrocarbon chains, partly saturated by the hydrogenation.

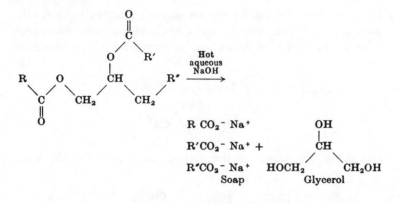

Sodium chloride is added to the aqueous solution containing the soap and glycerol. The soap is precipitated and filtered off and refined by washing, adding perfume and finally formed into tablets. The glycerol is recovered from the filtrate by concentration of the aqueous solution.

(b)

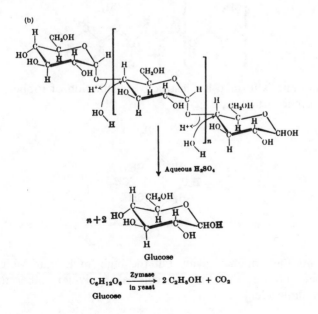

Glucose

$$C_6H_{12}O_6 \xrightarrow[\text{in yeast}]{\text{Zymase}} 2\ C_2H_5OH + CO_2$$
Glucose

2. Pentan-4-onal:

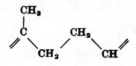

the formation of which by the ozonolysis of rubber indicates the presence of double bonds, four carbon atoms away from each other in the rubber molecule, i.e. a structure having a repeating unit:

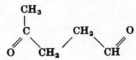

and one double bond per five carbon atoms joined in this way:

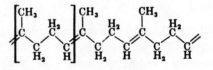

rather than

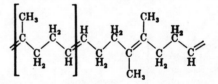

as isoprene is formed by destructive distillation of rubber.
A cyclic structure, e.g.

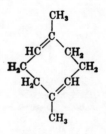

also fits the evidence given, but the high 'molecular' weight of rubber indicates that the molecule is a polymer rather than a cyclic molecule.

3. (a)

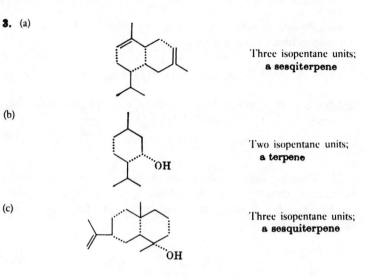

Three isopentane units;
a sesqiterpene

(b)

Two isopentane units;
a terpene

(c)

Three isopentane units;
a sesquiterpene

Chapter 18

1. (a), (b), (c) Tetraethyl lead, poly(ethene), ethanol and ethanoic acid may all be prepared from ethene.

Ethene is obtained either from natural gas which contains ethane by thermal decomposition (cracking):

$$CH_3CH_3 \xrightarrow[1000°]{\text{Heat}} CH_3\dot{C}H_2 + H\cdot$$

$$CH_3\dot{C}H_2 \longrightarrow CH_2CH_2 + H\cdot$$

or as a by-product in the cracking of higher hydrocarbons:

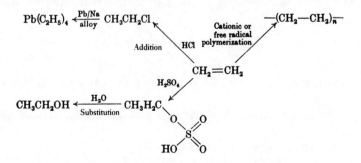

(e), (f) Poly(vinyl chloride) and ethanoic acid may be obtained from ethyne, which, like ethene, is obtained by the thermal decomposition of natural gas:

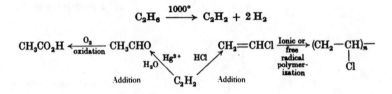

$$C_2H_6 \xrightarrow{1000°} C_2H_2 + 2H_2$$

(d) Toluene may be prepared from a fraction of petroleum containing about seven carbon atoms by a thermal decomposition reaction which dehydrogenates and cyclizes the heptane molecules in the C_7 fraction:

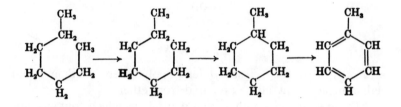

2. (a) Cyclohexane from reforming serves as a starting material for Nylon 66 (see page 224).

(b) p-Xylene, also obtained by alkane reforming, is converted into Terylene by the steps shown on page 227. (The ethane-1,2-diol used can most conveniently be obtained from ethene, i.e. from petroleum).

(c) Bakelite is a condensation polymer of phenol and methanal. Methane could serve as a source of methanal by controlled oxidation:

$$CH_4 + O_2 \longrightarrow CH_2{=}O + H_2O$$

(In Britain most methanal is made by the catalytic oxidation of methanol.)

The condensation of phenol and methanal is depicted on page 228.

(d) An azo dye. Starting materials:

Benzene   From reforming
Phenol   of petroleum
               alkanes

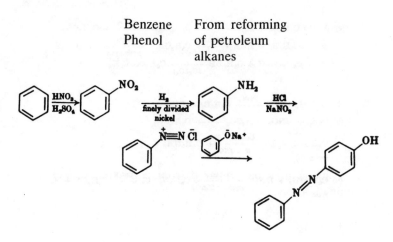

Chapter 19

1. (a) ⌐Either $3 C_2H_5OH + PBr_3 \longrightarrow 3 C_2H_5Br + H_3PO_3$

   or $C_2H_5OH + Na^+Br^- + H_2SO_4 \longrightarrow$
   $$C_2H_5Br + Na^+ + H_3O^+ + HSO_4^-$$

(b)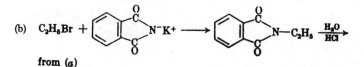

   from (a)

   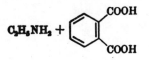

(c) $C_2H_5OH + Na \longrightarrow C_2H_5O^- Na^+ + \frac{1}{2} H_2$
   $C_2H_5O^-Na^+ + C_2H_5Br \longrightarrow (C_2H_5)_2O + Na^+ Br^-$
                    from (a)

(d) ⌐$C_2H_5Br + Mg \xrightarrow[\text{ether}]{\text{Dry}} C_2H_5Mg Br$
   from (a)

   $C_2H_5MgBr + CH_2{=}O \longrightarrow C_2H_5{-}CH_2{-}OMgBr \xrightarrow[\text{acid}]{\text{Dilute}}$
   $$C_2H_5CH_2OH$$

(e) Either $C_2H_5Br + Mg \longrightarrow C_2H_5MgBr \xrightarrow{CO_2} C_2H_5CO_2H$

or $\quad C_2H_5Br + KCN \xrightarrow{C_2H_5OH} C_2H_5CN \xrightarrow[\text{hydrolysis}]{\text{Acid}} C_2H_5CO_2H$

$\quad C_2H_5CO_2H + C_2H_5OH \xrightarrow{(H_2SO_4)} C_2H_5CO_2C_2H_5$

2. (a) $C_6H_6 + Br_2 \xrightarrow{FeBr_3} C_6H_5Br + HBr$

(b) $C_6H_6 + HNO_3 + H_2SO_4 \longrightarrow C_6H_5NO_2$

(c) $C_6H_5NO_2 + H_2 \xrightarrow[\text{or Sn/HCl}]{\text{Catalyst ,Ni}} C_6H_5NH_2$
from (b)

(d) $C_6H_5NH_2 + HNO_2 \xrightarrow[\text{aqueous } H_2SO_4]{\text{In dilute}} C_6H_5N_2{}^+ Cl^- \xrightarrow{CuCN} C_6H_5CN + N_2$
from (c)

3.

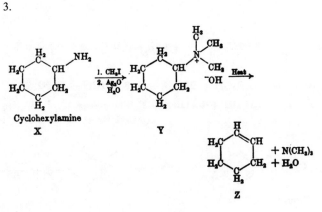

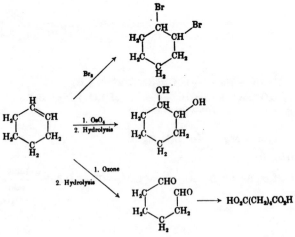

4. (a)

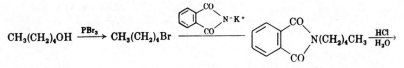

$$CH_3(CH_2)_4OH \xrightarrow{PBr_3} CH_3(CH_2)_4Br$$

$$CH_3(CH_2)_4NH_2 \text{ (as sequence for 1(b))}$$

(b) $CH_3(CH_2)_4OH \xrightarrow[\text{with } CrO_3]{\text{Careful oxidation}} CH_3(CH_2)_3CHO \xrightarrow[\text{2. Dilute acid}]{\text{1. } CH_3MgI}$

$$CH_3(CH_2)_3CH(CH_3)OH$$

(c) $CH_3(CH_2)_4OH \xrightarrow[\text{acid catalyst}]{\text{Heat with strong}} CH_3CH_2CH{=}CH_2 \xrightarrow[\text{2. Hydrolysis}]{\text{1. } OsO_4}$

$$CH_3CH_2CHOHCH_2OH$$

(d) $CH_3CH_2CH{=}CH_2 \xrightarrow[\substack{\text{2. Oxidative} \\ \text{hydrolysis}}]{\text{1. Ozone}} CH_3CH_2CO_2H + HO_2CH$

from (c)   Propanoic   Methanoic
            acid   acid

The more volatile formic acid may be separated from the propanoic acid by fractional distillation.

(e) $CH_3(CH_2)_3Br \xrightarrow[\substack{\text{in liquid} \\ \text{ammonia}}]{CH{\equiv}\overset{-}{C}Na^+} CH_3(CH_2)_3C{\equiv}CH \xrightarrow[\substack{\text{2. 1 mole} \\ CH_3(CH_2)_3Br}]{\substack{\text{1. } NaNH_2 \text{ in} \\ \text{liquid ammonia}}}$

from (a)

$$CH_3(CH_2)_3C{\equiv}C(CH_2)_3CH_3$$

5. The reaction of electron donors (X^-) may be generalized:

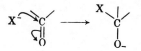

followed by

if the electron donor X^- was formed by the ionization of the compound XY.

(a)

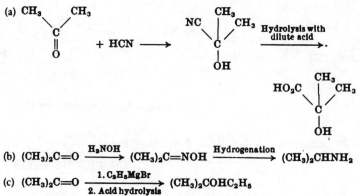

(b) $(CH_3)_2C=O$ $\xrightarrow{H_2NOH}$ $(CH_3)_2C=NOH$ $\xrightarrow{Hydrogenation}$ $(CH_3)_2CHNH_2$

(c) $(CH_3)_2C=O$ $\xrightarrow[\text{2. Acid hydrolysis}]{\text{1. } C_2H_5MgBr}$ $(CH_3)_2COHC_2H_5$

(d) $(CH_3)_2C=O$ $\xrightarrow{Reduce}$ $(CH_3)_2CHOH$ $\xrightarrow{PBr_3}$ $(CH_3)_2CHBr$

APPENDIX

Table A.1 Some characteristic infrared absorptions

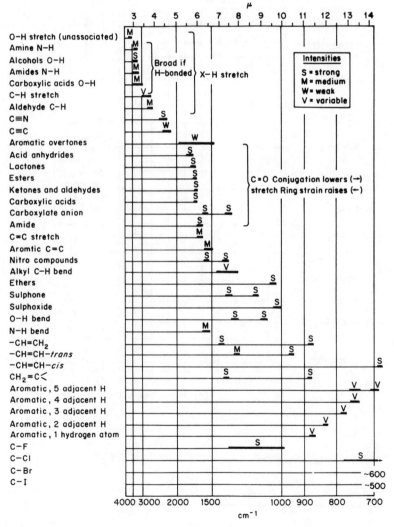

Table A.2 Characteristic chemical shifts of protons in different environments

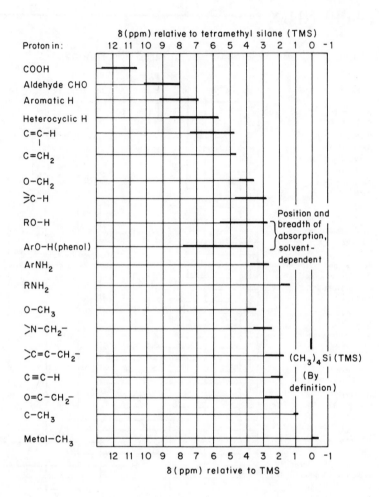

Table A.3 Characteristic chemical shifts of ^{13}C in different environments

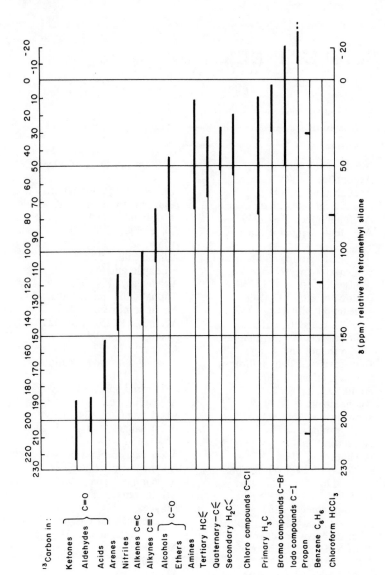

SUBJECT INDEX